UFO SCIENCE

SECRET NEW PHYSICS, VEHICLES, AND UAP

G O TURNER

EDITED BY
KAZ MORRAN

CONTENTS

ALSO BY THE AUTHOR

PUBLISHED BY A LITTLE VAST STUDIOS

The Books of Ruein

Prequel Novelette:**Rue's Requiem**

Novelette:**Mothers Lament**

AVAILABLE AS FREE EBOOKS ON GOTURNERWRITES.COM

———

Book One:**The Book of Ruein**

———

HARADEN TRILOGY

Book Two:**Fires of Haraden**

Book Three:**Nazier's Visage**

Book Four:**Gilded Lands & Cages**

PUBLISHED BY A LITTLE VAST STUDIOS LLC

Inquiries may be addressed via email to: social@goturnerwrites.com.

First Edition: October 2023
ISBN 979-8-9891459-0-4
Library of Congress Control Number: 2023946416

Hey, not to be pitchy or anything, but if you found this guidebook to be mind blowing, then I invite ya to sign on to my mailing list on my website. That way you'll get the occasional notice when future publications are released, as well as access to exclusive contests, giveaways, and freebies.

Your email will never be shared. Unsubscribe at any time.

INTRODUCTION

For decades, the mysterious phenomena of unidentified flying objects, or UFOs, have captivated people. We've spotted these strange and advanced aircraft in the skies all around the world. They often exhibit behavior that defies the laws of physics as we know them. The topic has gained renewed interest with the 2017 Department of Defense (DoD) disclosures appearing in the *New York Times*. This has given rise to several podcasts and cable news features. All of this has led to an abundance of speculation about the phenomena.

That is where this book comes in.

Set aside the tinfoil hats for a bit, and screw on your thinking caps for some thought-provoking concepts.

Our goal is to gather the latest and most promising theories about UFO physics in one comprehensive guidebook. By compiling and examining the evidence driving each craft, we aim to provide a clear overview of the current state of our understanding. Gathered here, we have the claims from scientists like Salvatore Pais to the controversial Bob Lazar. There are chapters on alien reproduction vehicles reportedly developed by a US military contractor, as well as case studies built around declassified Pentagon incidents.

**Rather than filling every other paragraph with adverbs like suppos-
edly, allegedly, purportedly, etc., each chapter is presented under a
premise of truthful testimony. You're invited to review, consider a
range of possibilities, and examine the hypotheses for each.**

Whether you're a UFO enthusiast looking to deepen your knowledge,
a scientist open to explore new theories, or simply someone filled with
curiosity, this guidebook offers a wealth of mind-bending ideas
combined with a real-world understanding. Join us as we delve into
the unknown and consider the implications for our place within the
universe.

CHAPTER 1
BELIEF OR REALITY

WE'VE all seen the news features. They open with the *X-Files* theme music and ask, "Do you believe...?" It's unceasingly delivered with tongue firmly planted in cheek. Their obvious intent is to preserve the seriousness of a "news" organization by blowing an air of ridicule on the topic.

Now, think about this. We already know there are billions of galaxies within the observable universe. Recent surveys have revealed the vast majority of stars have planets of their own. It is becoming clearer with each passing year that the likelihood of there being other habitable worlds is overwhelming.

Yet, similarly to how our brains vie between logic and emotion, our modern society's view is of two minds when it comes to the subject of unidentified anomalous phenomena (UAPs). An intellectual understanding that life must exist beyond us is clear, but our emotional reaction is to vehemently reject the notion at the same time.

A recent Pew Research study concluded, "About two-thirds of Americans (65%) say their best guess is that intelligent life exists on other planets." And yet, if the topic is brought up individually, oddball glances are the norm.

What? Are you some nut job?

LET'S DISPEL THE "BELIEF"

We'll begin with December 16, 2017 and the *New York Times* front-page headline, "Glowing Auras and 'Black Money': The Pentagon's Mysterious UFO Program." This feature was submitted by journalists Leslie Kean and Ralph Blumenthal. In the article, the DoD revealed they were funding an Advanced Aerospace Threat Identification Program (AATIP), regarding unidentified flying objects from 2007 to 2012. A follow-up article, "2 Navy Airmen and an Object That 'Accelerated Like Nothing I've Ever Seen'" highlighted a newly declassified account of one of the most compelling incidents from 2004.

Prior to AATIP, the United States military conducted a series of studies on UFO sightings from 1947 to 1969, the most well-known being Project Blue Book. While initially claiming to have thoroughly investigated and explained most sightings, it has since been revealed the Project's main purpose was as a disinformation campaign. This was done out of fear of Communist forces exploiting the phenomena to instill paranoia in open society. As a result, the goal became to dismiss the topic and foster the very stigma we live with today.

The DoD's more modern replacement, AATIP, was primarily funded at the request of Harry Reid, a former Senate majority leader. The program was headed up by military intelligence official Luis Elizondo. AATIP collected video and audio recordings of reported UFO incidents. The 2015 footage from a Navy F/A-18 Super Hornet showing an aircraft moving at high speed and rotating (the Gimbal video) was an example of this.

Eric W. Davis, an astrophysicist who has worked as a subcontractor and consultant for the Pentagon's UFO program, opened up about physical materials being studied. These pieces could not be identified as human-made. Mr. Davis also confessed to giving a classified briefing to a DoD agency about "off-world vehicles not made on this earth."

Since AATIP's funding was discontinued, new programs have stepped up and superseded. The Unidentified Aerial Phenomena Task Force (UAPTF) was a program within the Office of Naval Intelligence responsible for collecting and reporting sightings. In 2021, the UAPTF released a preliminary report, which evolved into the creation of the All-domain Anomaly Resolution Office (AARO).

The crux of what I'm getting at is the US government is no longer avoiding the phenomena. There is something in our skies. It is physically present, and currently offered to the public as "unknown."

Indeed, as of June 5, 2023, a highly qualified Air Force intelligence officer stepped down to become a whistleblower. David Grusch is a decorated combat officer, veteran of the National Geospatial-Intelligence Agency (NGA), and was the representative of the National Reconnaissance Office (NRO) to the UAPTF from late 2021 to July 2022. He has filed a claim with the U.S. Office of the Intelligence Community Inspector General (ICIG) asserting shocking allegations.

Grusch has testified under oath to the ICIG and been interviewed by the same Leslie Kean and Ralph Blumenthal from the *New York Time's* articles. David asserts the US federal government maintains a highly secretive UFO retrieval program and possesses multiple spacecraft of non-human origin, as well as corpses of deceased pilots.

And yet, there is hardly any coverage of this from mainstream media outlets.

Take a moment to digest that.

Despite it being over five years since the DoD made their initial disclosure; the stigmatization is as strong as ever. Today, there remain two Great Mysteries: *Is there life after death?* and *Are we alone?* Isn't it about time we moved past the ridicule? Or are we going to allow it to keep us from resolving one of the greatest unsolved questions in human history?

Even physicist Dr. Michio Kaku argues, "The burden of proof has shifted. It used to be the burden was on the people who believe in UFOs. Now the burden of proof has shifted to the Pentagon, to the military. Now they have to prove that these aren't extraterrestrial."

The sun rises in the east. The moon is present even when you do not see it. And water is wet. If, like Einstein, you take a materialist view of the universe, then reality does not require belief.

CHAPTER 2
WHY CAN'T WE?

YOU MIGHT ASK, what's the big deal with this subject? We've got fighter jets that break the sound barrier and rockets that blast us to space and the moon. We have probes that have ventured beyond our solar system. Why aren't we just choosing to visit other worlds?

Isn't this just a question of speed and distance?

Actually, yes.

Undoubtedly, you've heard of a little thing called the speed of light. This is not just really fast, and it's more than a speed limit. You literally cannot achieve this speed. And even if you could, it still wouldn't be enough. Here's why:

1. You and all things physical (solid, liquid, gas, plasma) are made of baryonic particles (up and down quarks orbited by electrons). These baryons form atoms, also more commonly known as matter.

2. All matter has weight because of an interaction with the Higgs field. What's a Higgs field? *I'm going to give an oversimplification here to make something very complex digestible.* Visualize an ocean that all the universe resides in. As matter crosses through this Higgs ocean, it meets resistance or drag as it moves. Higgs bosons get picked up and conveyed by

matter. This is a rough way of understanding how weight imbues all physical material.

3. The photon is understood to be the exception. Photons are quantized packets of light, and do not interact with the Higgs field. Photons are weightless. As such, they have a top speed of 186,282 miles (299,792 km) per second. Sounds fast, right? And it is. At that speed, one could travel from the sun to Earth in eight minutes. However…

4. Light speed is slow on a universal scale. Our closest neighboring star is four light-years away. That literally means it takes a ray of light four years to travel that distance. Our Milky Way galaxy is between 150 to 200 thousand light-years across. The Andromeda galaxy is 2.5 million light-years away.

5. Now, here's the bad news: Because we're made of matter and interact with the Higgs field, the faster we go the heavier we get. The heavier we get, the more energy is required to move us faster. This weight gain becomes exponential the closer you get to the speed of light. The amount of mass you gain at light-speed would amount to infinity. That means you'd weigh as much as all the universe, all at once.

So, *not possible*. And I haven't even brought up time dilation effects yet!

This is the crux of why most modern arguments say that interstellar travel is just not feasible. Einstein irrevocably asserted that *nothing* travels faster than light. Even light can't go faster than light. It's an important distinction that every intellectual just needs to accept. The science is settled. There's just no getting around it.

Right?

There have been moments throughout history when the science has been settled, only to be overturned later. Four hundred years ago, Earth was the center of the universe. Back then, all the heavens orbited around us. Then Galileo Galilei plucked us out of our self-importance and declared that the solar system orbited the sun and not our planet. For his heresy, he was locked up for the rest of his days.

In 1886, Jules Verne helped coin the saying, *If God had intended for us to fly, he would have given us wings.* Yet a mere seventeen years later, the world was amazed by the Wright brothers' first powered flight in Kitty Hawk, North Carolina.

What all innovative thinkers have encountered, since the dawn of science, is that the closed-minded guard their beliefs as ardently as Mother Nature keeps her secrets.

CAN THERE BE NEW PHYSICS?

Short Answer: There must be.

That's what every physicist has been working toward since our last advancements by Max Planck, Albert Einstein, Niels Bohr, Werner Heisenberg, and Erwin Schrödinger a hundred years ago. And the conundrum that vexed these great minds remains to this very day.

There appears to be two disparate sets of laws for existence; one centered on Einstein's general relativity (GR), which governs the macroscopic world we live in and perceive. Then there's a second set of laws that rule the atomic realm, called quantum mechanics (QM). But that can't be right. Both realities exist as our one reality. Add to that, their biggest point of contention is gravity and how unable they are to unify these two realities around it.

In truth, we've proven both GR and QM correct. So, in order to complete our understanding of physics, science needs to find some missing puzzle piece that bridges these two disparate laws of reality.

Most efforts have been focused on string theory. This framework suggests the universe's fundamental building blocks are not point-like particles, but extremely tiny objects known as strings. These multidimensional strings vibrate at different frequencies to produce the various particles and forces observed in nature.

POLARIZED QUANTUM VACUUM

Way down at the bottom of quantum mechanics, at the tiniest measurable distance (the Planck scale), the fabric of reality is still just spacetime. But, because of a quantum factor called the Heisenberg uncertainty principle, this empty space is not truly empty. Even in the nothingness of an interstellar void, random fluctuations in spacetime cause virtual particles to constantly pop in and out of existence. These particles are polarized, meaning they have a preferred direction of spin. We sometimes refer to this as quantum foam. It is a constant frothing of particles bubbling in and out of our reality.

As early as 1913, Albert Einstein and Otto Stern originated a concept we now call the quantum electrodynamic (QED) vacuum, or zero-point energy (ZPE). In 1948, Hendrik Casimir, a pioneer in superconductors, discovered that zero-point fluctuations in a photon field are affected by nearby conducting bodies. That means the vacuum's polarity can be changed externally.

How's all this quantum minutiae relevant to UFOs, you ask?

There is a commonality with most of the craft proposed in this book. It appears there is a connection between the Higgs field, zero-point energy, and how their interaction wends its way into our material world.

While no two chapters share identical takes on the subject, it seems clear there are tantalizing reasons to conduct further research.

————

We should touch on one additional topic.

The current understanding in the UFO zeitgeist is, if we have recovered non-human craft, the hardest part for our scientists to unravel is what we call material science. This focuses on comprehending and manipulating the properties, structures, and behavior of various substances. You see, new properties can arise from combining elements, forming alloys, and melding them into patterns, leading to novel outcomes.

Case in point, if you were to introduce a steel rod to an ancient

Egyptian pharaoh's court, they would quickly learn how incredibly strong it was compared to the basic iron they'd been working with. However, they'd have absolutely no idea how to forge such metal.

Now imagine you recovered something from a civilization at least a thousand years more advanced than we are.

METAMATERIALS

The majority of people today have never heard the term metamaterial. It's a relatively new field in material science (coined in 1999).

Metamaterial is defined as an engineered material designed to have properties not found in naturally occurring substances. It is created by arranging tiny structures or components precisely to manipulate and control the behavior of electromagnetic waves or other types of waves in unconventional ways.

You can relate it somewhat to waveguides. We use waveguides in everyday electronics and appliances to gather and redirect electromagnetic waves, such as in fiber-optic cables or microwave ovens. However, the difference is that a metamaterial acts more like a funhouse mirror or converting lens, warping EM wavelengths rather than simply to redirect.

Examples of what we've achieved:

1. **Electromagnetic cloaking:** These materials aim to redirect electromagnetic waves around an object, making it appear invisible to those waves. It can manipulate the path of light or other electromagnetic radiation, effectively cloaking the object within.

2. **Acoustic:** Designed to manipulate sound waves, these materials control the propagation of sound, bend or steer it, and even create "invisibility cloaks" for sound. Acoustic metamaterials have applications in noise control, ultrasound imaging, and underwater sonar systems.

3. **Negative index:** These materials exhibit a negative refractive index, which means they can bend light in unusual ways. They have been created using various designs, such as split-

ring resonators and fishnet structures, to achieve negative refraction.

4. **Superlens:** Designed to overcome the diffraction limit of conventional lenses, enabling imaging at sub-wavelength scales, these metamaterials can focus light or other forms of electromagnetic radiation beyond the diffraction limit, allowing for high-resolution imaging.

5. **Photonic band gap:** These materials exhibit a range of frequencies that cannot propagate through the material. They control and manipulate the flow of light, enabling applications such as efficient LEDs, optical fibers, and high-performance solar cells.

I will note that while these metamaterials have been successfully manufactured in laboratory settings, their practical applications are still being explored. Many challenges exist in scaling them up for widespread use.

The form of manipulation employed depends on what energy frequency you're looking to manipulate. The higher the frequency, the smaller its waveguide must be. Labs have been able to manufacture some things down to a microscopic scale.

However, some recovered material appears crafted on an atomic scale.

CHAPTER 3
UFOS | UAPS | ETC.

SOMEONE POINTS SKYWARD AND YELLS, "Look, a UFO!"
Ever notice how many people say UFO when they are referring to a flying saucer, cigar-shape, wedge, or some other craft? The term has become interchangeable even though it has a literal meaning.

UFOS

It is important to remember that the term UFO is an acronym for *unidentified flying object*. The U in UFO means *unknown*, so the object in question is a flying object that has not been identified or explained. This does not mean the object is extraterrestrial. It simply means the observer does not know what the object is. While many like to jump to the conclusion of an extraterrestrial origin, it is also possible that a variety of other factors, such as experimental aircraft, weather balloons, drones, or even optical illusions or hoaxes could cause the sighting.

Now look, I'm not saying to stop referring to the phenomena as UFOs. Given the diverse craziness of everything observed, and the vast range of shapes, sizes, and conditions, a catch-all term is clearly warranted. Just don't allow your critical mind to lose sight of what the

U stands for. This is just one helpful step toward promoting a more balanced and nuanced understanding of the subject.

UAPS

With the overwhelming success in stigmatizing the phenomena, the US government eventually found the old UFO acronym too charged a term to be used in serious circles. For multiple reasons, they adopted a new acronym, UAP: *unidentified anomalous phenomenon*. It is a more neutral term than UFO, creating an impartial distance from extraterrestrial origins.

This term remained in covert use until revealed on a talk show by a 2016 presidential contender. Now it has become a part of the topic's vernacular.

The military classifies UAP based on observations. There are several elements considered when determining the classification. These include the obvious — size, shape, color — and also:

- **Behavior:** How the object moves or behaves, such as whether it is stationary or in motion; and, if in motion, whether it is accelerating, decelerating, or maintaining a constant velocity; even the angle at which it moves.
- **Sound:** Any noise that may be associated with the object, such as engine roar/hum or other tones.
- **Duration:** The time that the object was observed.
- **Distance:** This is mainly regarding the span between the observer and the object, as well as the altitude.

These were all fine and good standard details. But what could we do to separate the frequent 'common' unknowns from the more intriguing UAPs?

THE FIVE OBSERVABLES

Former military intelligence officer, Luis Elizondo, has spoken about the importance of "the observables" in UAP research and analysis. He has emphasized the need to analyze and document in order to understand what an object and its origins might be. These observables are:

1. **Anti-gravity:** The ability of the object to defy the force of gravity and stay suspended in the air, lacking any flight surfaces.
2. **Sudden or Instantaneous Acceleration:** The ability of the object to move rapidly and unexpectedly, reaching ultra-high speeds in a short period, making right-angle turns, and even near-instant stops. Any of these maneuvers would seem lethal for occupants by our current inertial understanding.
3. **Hyper-speed Without Signatures:** The absence of "signatures" such as vapor trails and sonic booms that are typically left behind when an aircraft exceeds the speed of sound.
4. **Low Observability:** Our inability to detect the object visually, by radar, and other technological systems, or to rapidly change course to evade detection.
5. **Trans-medium:** The ability of the object to transition between different mediums, such as air, water, and space.

These observables are not necessarily seen in every UAP sighting, and it is not clear how commonly they are observed. It is unknown whether these observables indicate a particular type of UAP or are simply characteristics that have been observed in some instances.

OTHER OBSERVABLES?

Some craft appear to affect the physical and psychological well-being of people who have had close encounters. There have been cases of individuals reporting a range of harmful effects after a UAP encounter, including morphological changes to the body and brain.

For example, some have reported a sudden and intense fear or anxiety while in the presence of orbs, as well as physical symptoms such as nausea, dizziness, and difficulty breathing. Other individuals have reported unusual changes in their cognitive abilities or senses, such as difficulty concentrating. There are also physiological conditions, such as flash burns and hair loss.

While it is not known if a UAP causes these effects itself or by other factors, these are potentially significant observables that could provide insight into their nature and capabilities. We would need further research and analysis to understand the underlying causes and to determine their potential significance.

———

Now that we've established a basic footing and narrowed our focus, let's look at our first candidate.

CHAPTER 4
BACKSTORY: BOB LAZAR

IN 1989, KLAS 8's George Knapp conducted an exposé over several months on UFOs, during which he interviewed a silhouetted figure using the alias "Dennis." This unseen man's shocking revelation: the US government had employed him to reverse engineer extraterrestrial spacecraft. He'd performed the work at a top-secret facility known as S-4. Because of this broadcast, for the first time ever, the term Area 51 came into public knowledge.

The bombshell story quickly gained worldwide attention,

becoming a media sensation as far away as Japan. This controversial assertion captivated the imaginations of millions.

Months later, in desperation and backed into a corner, the shadowy whistleblower was convinced his only chance for survival was to step fully into the light. Thus, George Knapp introduced us to Bob Lazar.

This chapter is based on his account.

Rather than rehashing the details already accounted in several other books and documentaries, I'll hit the major bullet points. Along the way, however, I'll pepper in a few additional details learned.

Researchers are at a loss with finding Robert Scott Lazar's educational records. On paper, this man is an electronics and physics engineer without a degree. He seems to have never entered higher academia.

This brings us to our first tidbit. Respected ufologist, Linda Moulton Howe, opened up about her private discussions with Bob. She shares that part of the issue is that Bob attended some universities not as a matriculated student, but as a free rider. Unable to afford the degree, Bob rode in on the coattails of friends. He'd sit quietly in the back taking in all he could.

Regardless, he somehow ended up landing an entry position at Los Alamos Meson Physics Facility in 1982. *The Los Alamos Monitor* corroborates this in an article headlined, "LA man joins the jet set—at 200 miles an hour," featuring Bob's fun side.

Source: Los Alamos Monitor *1982*

In 1986, after marrying his second wife, Tracy Ann Murk, Bob ran a film processing business in Las Vegas. He also started up an annual festival with his good friend, Gene Huff, that would develop into "Desert Blast."

Through Gene, Bob was introduced to John Lear in 1988. John Lear was an accomplished test pilot and son of the Learjet magnate, Bill Lear. Now, Bob characterized John as a wild man, and one happy to go along with just about any conspiracy. It was John Lear who first told Bob about Area 51 and how they have alien craft that were being back engineered. Bob was initially skeptical. Still, John's reams of collected UFO documents intrigued him all the same. He offered to trade setting up equipment in return for getting copies of his UFO papers.

That's when Bob happened upon Dr. Edward Teller, one of the fathers of the hydrogen bomb. This respected figure was at Los Alamos for a lecture. Bob found him reading the local paper, which included his rocket car story in it. Impressed, Dr. Teller struck up a conversation. Bob told Dr. Teller he wanted to be doing the sort of work they do out at Area 51.

With Dr. Teller's recommendation, he landed that position with Edgerton, Germeshausen, and Grier, Inc. (EG&G). After a lengthy background check, he was determined to be stable enough and without questionable connections. Bob accepted the sensitive nature of what they were hiring him for. He also agreed to be monitored to ensure government secrets were kept.

After boarding a Janet Airlines flight from McCarran Airport, Bob arrived at Area 51. He was introduced to his base supervisor, Dennis, and driven in a blacked-out bus several miles south to a second, more covert base.

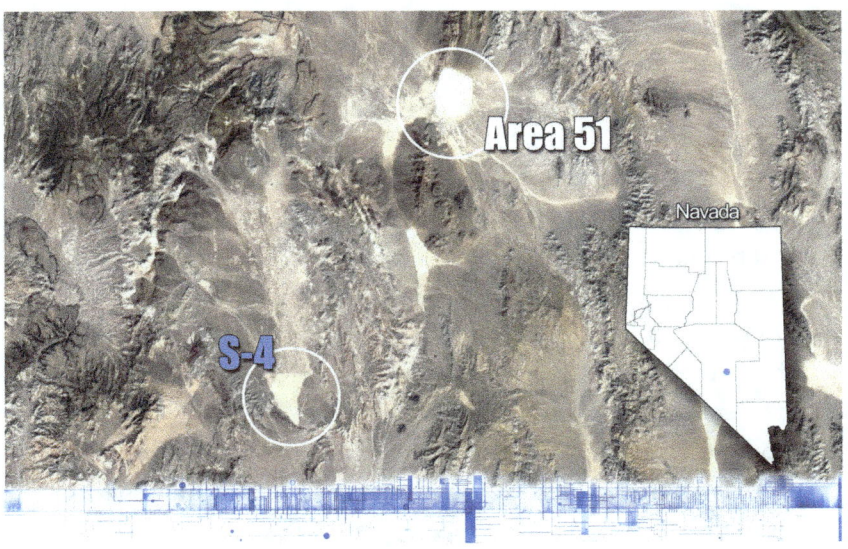

S-4

They led Bob to a private room, where he read a brief. It was incredibly detailed with everything one might expect from a UFO research project. Still skeptical, Bob thought he was being tested, unsure if he was reading a load of malarkey. It was all just too fantastic.

He met his research partner, Barry. After a startling lab demonstration, they led Bob out into a hanger. That's where he saw the Sport Model for the first time. It was a classic flying saucer, but with an American flag decal on it. Bob's first thought: this explained all those UFO sightings. He was probably looking at some new top-secret prototype from EG&G.

Only once he realized the propulsion being generated and the size of the craft's interior did he fully understand... human hands did not make this technology.

Between December 1988 and March 1989, Bob had an on/off schedule at the S-4 facility. Often, he'd receive phone calls at odd hours, such as at 11 PM along with the expectation that he be on a flight by 11:45 that same night. He'd work for long stints before returning home.

For Bob, this was some of the most exciting but ominous work he would ever do. You can understand why there'd be little room in his mind for anything but this.

That is, until it stopped.

SHUT OUT

After those heady first three or four months, Bob didn't get any more calls. A week turned into two — but still no call to return. It baffled him to be given such an astounding prize and then have it suddenly withheld in silence. Bob wondered about what he'd been shown. Worry crept in. Why weren't they calling?

Would they just let him walk away knowing what he knows?

Since it was Gene Huff and John Lear who started him on this path, he decided he needed a just-in-case plan. Bringing his wife in on it, Bob opened up to them about everything. To prove his case, Bob also knew S-4's test-flight schedule. They could see for themselves.

Every Wednesday night for the next three weeks, they took an RV into the desert and witnessed a glowing object perform impossible maneuvers over the base. Getting cockier with each trip, on that third night, they got caught.

The following morning, Dennis brought Bob in for interrogation. There'd be no reasoning out of this. Bob had crossed a serious line. Hours in, over yelling and brandished weapons, Dennis revealed why the lab work had stopped coming.

All the late-night calls, since Bob became part of the program, had eroded the trust in his and Tracy's relationship. With Bob leaving for days without disclosing what he was up to, Tracy thought Bob was cheating. So she began an affair of her own. Since Bob and Tracy's phone lines were tapped, intelligence officers found out before Bob and suspended his clearance. A wait-and-see policy began based on how his marriage would shake out.

Concluding the interrogation with a series of threats to Bob, his wife, and his friends, they finally released Bob. Then a system of erasure moved into place. Much of Bob's history vanished overnight. Gone were his birth certificate, academic degrees, and employment history. If not for IRS filings with the Department of the Navy, there'd be nothing to show of any government work.

That May, while Bob was entering a freeway on-ramp, a car raced up on the berm beside him. He heard a gunshot. Bob veered, over-corrected, and ran off the road while the other car sped away. A passenger-side tire had been shot.

At this point, what would you do?

KLAS EXPOSÉ

The threat on Bob's life had become all too real. Once again turning to John and Gene, Bob was advised to strike back in order to protect himself. If he could get word of what was happening out publicly, then the government wouldn't dare kill him. So, Bob turned to KLAS and conducted George Knapp's silhouetted interview. As an added thumb in the eye of his supervisor, Bob used Dennis' name as his alias.

In some respects, it worked. But since Bob had become untouchable, the forces arrayed against him moved on to anyone associated with him. Distant relatives, acquaintances, and friends abruptly began having governmental administrative problems. Bob was still monitored and followed. His car and home were broken into.

Fed up with all the drama, Bob revealed himself on camera that November. In the thirty-plus years since, Bob's account has remained steadfast, and many of his outlandish claims have been verified.

Now let's dig into what he learned.

CHAPTER 5
BOB LAZAR'S SPORT MODEL

WHILE AT THE BASE, Lazar observed a variety of spacecraft (nine in total), some of which he coined terms for, like "Jell-O Mold" and "Straw Hat." According to Bob, all nine craft functioned under the same form of technology. However, the only vehicle Bob was permitted to examine directly, he nicknamed the "Sport Model."

In order to preserve secrecy and not allow any one person to know everything, work was compartmentalized into three projects. The base commanders paired up various experts, with each group focused only on their aspect of the vehicle.

- **Project Galileo** dealt with gravity propulsion and was Bob's selection.
- **Project Sidekick** examined a beam weapon, focused by a gravity lens.
- **Project Looking Glass** was assigned to work out the physics relating to seeing backward in time.

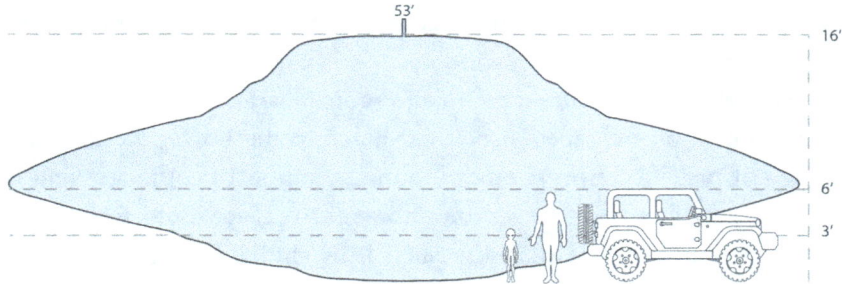

OVERALL APPEARANCE AND LAYOUT

This is your classic flying saucer, measuring nearly 53 feet (16 m) wide and 16 feet (4.9 m) tall. Bob describes the exterior as brushed-metal, forming a dull aluminum/pewter-like finish. It is unclear what substance the ship's creators crafted it from or whether it's pure metal, an alloy, or (most likely) a metamaterial. Both the exterior and interior are colorless, exhibiting no seams, rivets, or bolts of any sort. Apparently devoid of right-angles, the entire vessel seems to be formed from a mold or 3D printed.

> "It was beyond being rounded; it was like it was almost melted. It looked like it's made of wax and heated for a time and then cooled off. Everything has a soft round edge to it; there's no abrupt change in anything. It looked like everything was cast out of one piece."[1]

The three levels of the interior are divided into a Lower Deck, a Main Deck, and an Upper Deck, which is speculated to be a computer/navigation array. Because of Bob's compartmentalization to gravity propulsion, it is believed that the upper section did not pertain to Bob's team, and thus was kept closed to them. The Main Deck serves as the hub of the vessel, containing the reactor, amplifiers, and seats.

It is important to note that the design of these features was not optimized for adult human use. When Bob could go inside, he found he could only stand upright in the center of the craft. The consoles and seats were too small and low to the floor. It was as if they were made for small children.

Surrounding the hull's interior is a series of arches. When the craft was activated for Bob, one arch became transparent, providing a view of the exterior as if it were a window.

The Main Deck's floor had a hexagonal pattern and contained a collapsible, honeycombed-structured hatch to the Lower Deck. When closed, it provided firm enough footing to support people crossing it. However, with a finger insertion and slight tug, it would retract and fold upon itself easily. Bob described it as similar to cardboard beer bottle partitions. He also joked that of everything he witnessed, only that did he fully understand.

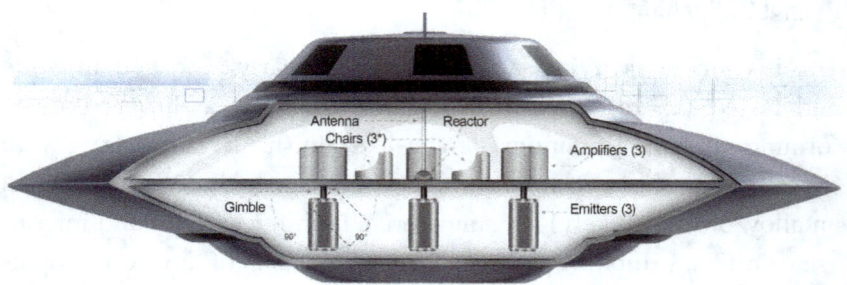

*Only two chairs are represented so the view of the reactor is not blocked.

ENGINE STRUCTURE

Centered on the Main Deck is the power plant, contained within a basketball-sized hemisphere. This acts not only as a gravity wave generator, but also as an anti-matter reactor. Connected atop it is a thin, transparent antenna/waveguide that extends through the top of the vessel. Equilaterally positioned from the deck's center are three gravity amplifiers. These are routed through to the Lower Deck and to the gravity emitters directly below each.

SO, HOW DOES IT WORK?

Essentially, the Sport Model operates similarly to an Alcubierre drive [*alku 'βjere*]. This is a warp drive concept; contracting space in front and expanding space behind it. Proposed by theoretical physicist Miguel Alcubierre in 1994, the drive offers a workaround to Einstein's field equations, which states objects cannot accelerate to the speed of light. Instead, the Alcubierre drive warps the fabric of spacetime around an object. This expansion/contraction essentially shortens the local distance, allowing the object to arrive at its destination more quickly than light would, while breaking no physical laws.

To achieve this, Alcubierre's concept requires a theoretical exotic matter that could produce negative energy density (anti-gravity). In our case, the Sport Model's fuel source, element 115, fulfills that role. However, rather than providing anti-gravity, this element produces gravity. Its A-wave generation can be phase shifted with the Earth's B-waves, allowing for both push and pull capabilities.

> "The amplifiers always run at 100%. They are always outputting a maximum gravity wave, and that wave is phase-shifted from zero to 180°. That's essentially the attraction and repulsion, and it's normally at a null setting somewhere in between. It's a very straight-forward system. It looks more like a coal-fired engine than very hi-tech."[2]

Gravity waves are ripples in the fabric of spacetime. These waves can have a wide range of frequencies, depending on the characteristics of the objects that are creating them and the strength of their gravitational interaction. They can become noticeable when two massive objects are in proximity to each other and interact through their pull.

An example of this was detected with LIGO (the Laser Interferometer Gravitational-Wave Observatory) when two black holes collided and merged.

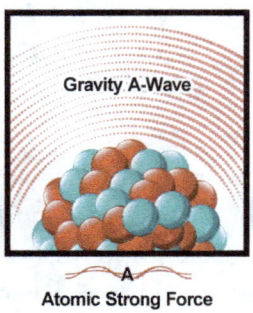

Gravity A-Wave

A

Atomic Strong Force

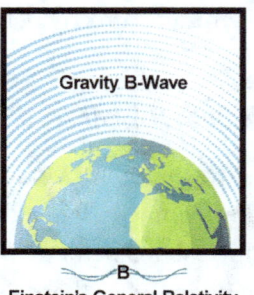

Gravity B-Wave

B

Einstein's General Relativity

According to Bob's understanding, there are two types of gravity: Einstein's General Relativity, where gravity results from large sums of mass (gravity B), and a quantum level form (gravity A). The latter version is what we know as the strong nuclear force. The standard model sees the strong nuclear force as a fundamental interaction that confines quarks into forming protons, neutrons, and other hadron particles. It also binds neutrons and protons to create atomic nuclei. It is a force that historically has been accepted as only existing on an atomic scale.

Bob has been led to understand that the strong force can extend itself outwardly from within the heaviest elements. This is the gravity A-wave. It generates as a continuous signal at a fundamental microwave frequency. That creates a wavelength extension of roughly an inch (26.2 mm) from the nucleus, where it becomes accessible.

While an inch doesn't sound like it would be of much practical use, for something originating on an atomic scale, it is a vast reach. Imagine an atom enlarged to the size of a basketball. At that scale, our original inch would now stretch well past the size of our solar system, all the way to the surrounding Oort cloud. And that's just comparing to a single atom. That atom's nucleus would be a floating speck inside the ball, one thousandth the size of its outer skin.

ELEMENT 115

115
Mc
Moscovium
290.196

This is where our heavy element 115 comes in. Not only the fuel source for our craft, it primarily provides a serviceable level of A-waves from its nucleus. This heavy atom had been theoretical until 2003, when it was manufactured within a particle accelerator at the Joint Institute for Nuclear Research in Dubna, Russia. Today, it is on the periodic chart as Moscovium 115. Because of the unstable nature of the isotope produced, it exists for only a fraction of a second before decaying into a lower state. Therefore, debunkers often state it isn't viable.

This is a false argument that is quickly refuted by another example element. Imagine you wanted to use a particle collider to manufacture atoms of gold. This *could* be done. Forget about how extremely expensive this method would be. Instead, understand that gold has forty-one known isotopes, ranging from gold-170 to gold-210. Only one of these, gold-197, is stable; the rest are radioactive. So, you'd have a one in forty-one scatter-shot chance of manufacturing a single, stable element of gold through this method.

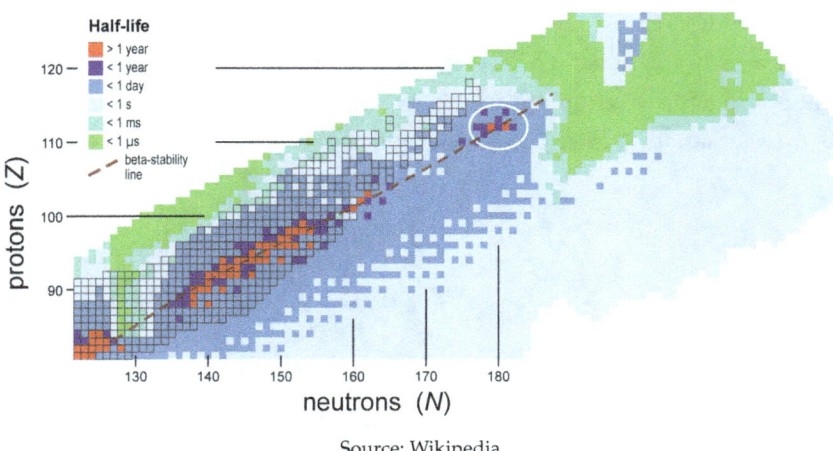

Source: Wikipedia

Contrary to the instability narrative, there is (accepted as a likelihood) an "island of stability" within Moscovium's reach. It would range from element 112 out through 115, where, if a balanced number of neutrons were present, the element could stabilize.

Yeah, but… considering how difficult this is to manufacture, how could there even be so much of this rare metal? It's not occurring naturally on Earth.

Fair enough. However, our solar neighborhood isn't as dramatic as others. Bob suggests element 115 might be native to systems which have been seeded by more violent stellar life/death cycles. Greater super nova explosions would generate higher density elements than ones manufactured by mid-range common stars like ours.

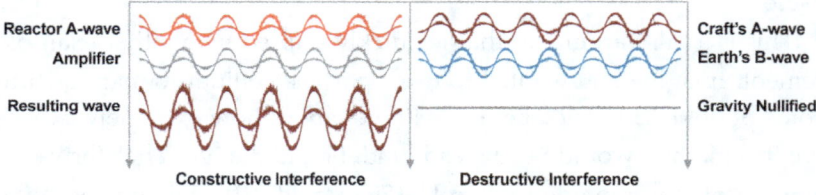

Now, once you can access the gravity A-wave, you can amplify it just like any other electromagnetic wave. The antenna's waveguide siphons off the wave, redirecting it around to the gravity amplifiers. The amplifiers boost the A-wave to the emitters, sequentially releasing them at a rapid pulse of 7.46 Hz. Control variables include changing the phase and direction of the A-wave against the phase of the gravity B-wave emanating from the Earth, thus negating gravity.

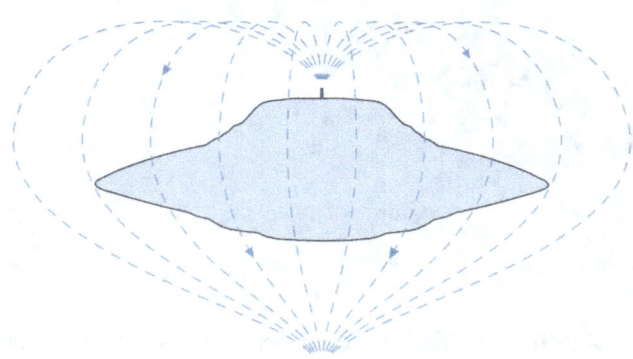

This amplified A-wave is redirected to not only envelop the craft, but also provide propulsion.

ANTENNA

Since the gravity A-wave operates in a microwave frequency, a wave-guide can siphon it off. A waveguide, for us humans, is normally a hollow copper or aluminum tube designed to redirect energies from one point to another. Its shape and size are precisely defined to ensure energy propagates efficiently.

REACTOR

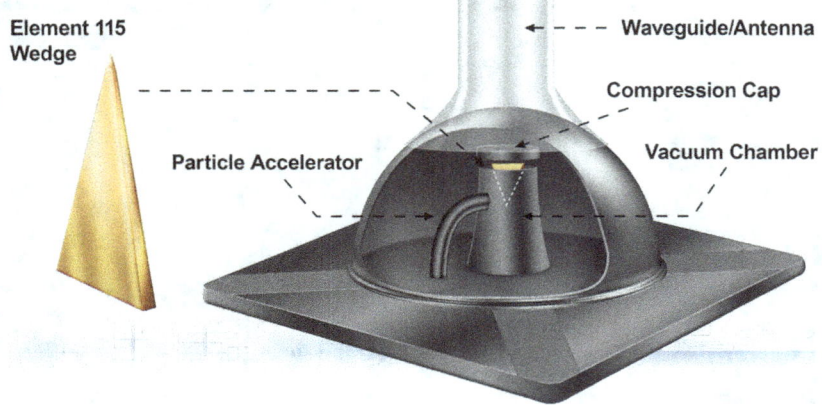

To power our flying saucer, we insert a wedge of element 115 into the top of a vacuum tube, centered on a base plate. We activate the reactor when its hemisphere is seated atop.

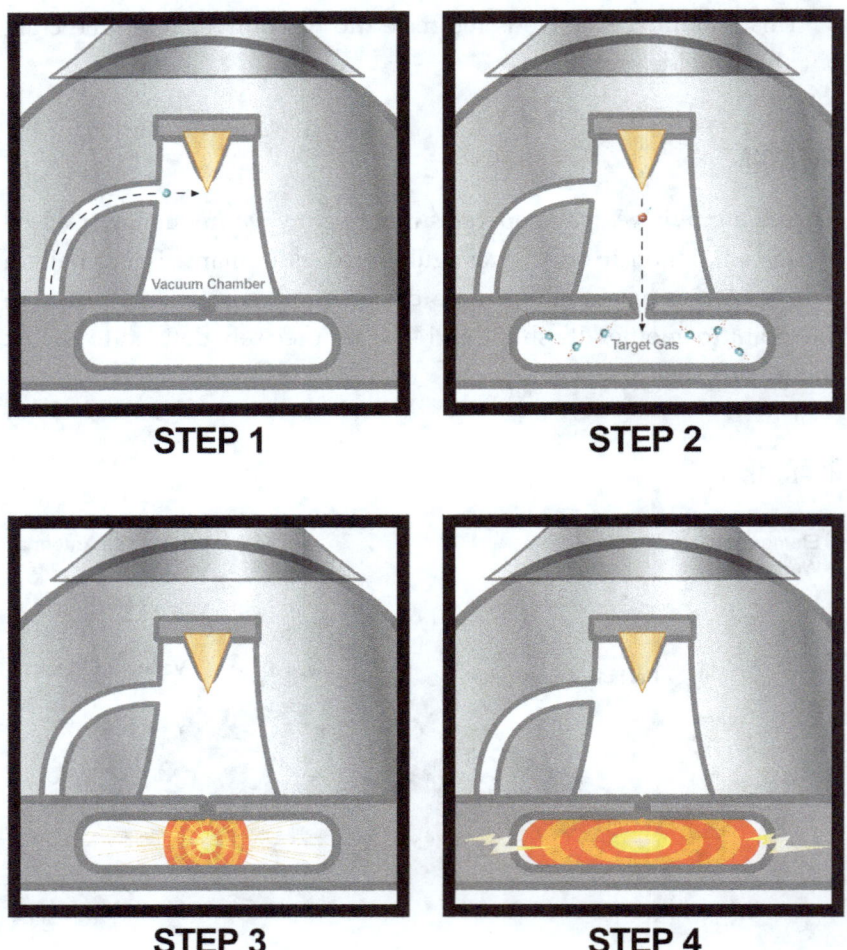

STEP 1

STEP 2

STEP 3

STEP 4

[**Step 1**] A particle accelerator fires a proton into the 115, fusing the bottommost atom into element 116. [**Step 2**] It spontaneously decays, releasing a small amount of antimatter. That antihydrogen is directed into a lower chamber, presumably by a magnetic field. Once sealed inside, hydrogen is released. [**Step 3**] The interaction of matter/anti-matter annihilates both, releasing 100% of their combined potential energy. [**Step 4**] This energy converts to electricity. Now remember, this craft has no wires or connections we'd be familiar with. Bob's best guess? That electricity is conducted/distributed through a process similar to a Tesla coil, in order to power the craft.

"All of these actions and reactions inside of the reactor are orchestrated perfectly, like a tiny little ballet. And in this manner, the reactor provides an enormous amount of power."[3]

The amount of power a matter/antimatter reaction generates is gigantic. Do the other instruments require such a huge resource? Or does the reactor only fire periodically when reserves are depleted? Either way, the power generated appears to be secondary to the craft's operation. Our flying saucer's primary driver is gravity.

The amount of 115 required is only half a pound (223 g) and has an expected utilization of approximately twenty to thirty years. According to Bob, the USAF has as much as five hundred pounds (227 kg) in their possession.

GRAVITY EMITTERS

OMICRON CONFIGURATION

Non-hinged rods act as gimbals, connecting the Main Deck's three amplifiers above to the Lower Deck's three emitters. These are directable up to a 90° angle from vertical, and a full 360° horizontally swivel. When the craft is maneuvering within a planet's atmosphere, the emitters take on an omicron configuration. Two emitters direct their A-waves down to negate the B-wave of the planet. That weightlessness allows for the remaining emitter to push/pull the ship in any other direction. Possibly because of this off-center alignment, the craft's performance at low speed often presents with a wobble.

Before initially lifting off the ground, the disc's underbelly briefly gives off a blue glow and coronal discharge, a sound similar to that of high-voltage electricity. Once airborne, it is completely silent, and the glow fades away.

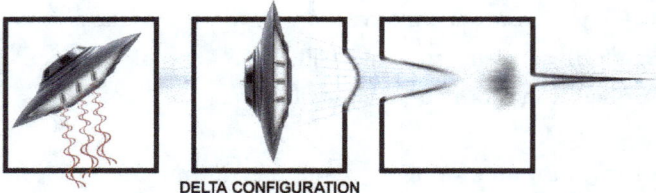

DELTA CONFIGURATION

For traveling vast distances, the emitters take on the delta configuration. The craft orients its belly in the intended direction, and all three emitters focus on a single distant point beneath, relative to the craft. The amplitude is brought to full power, and the Sport Model warps belly-first to its destination. It's like creating a bottomless black hole, but with an on/off switch.

Also consider the shape of the lower half of the craft. Bob believes that nothing on this vehicle was there without a reason. It seemed the beveled rings beneath acted as a gravity lens for the A-wave being emitted. The hull is likely an advanced meta-material, one capable of focusing gravity waves as well as storing power and redistributing the energy where needed.

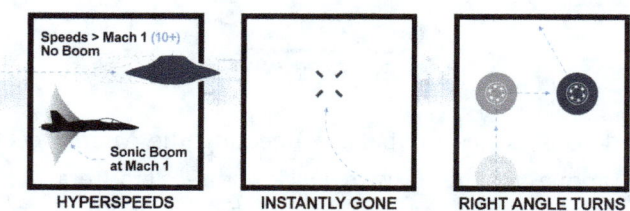

HYPERSPEEDS   INSTANTLY GONE   RIGHT ANGLE TURNS

Essentially, the drive creates a pocket of spacetime independent of its surroundings. Occupants become weightless. As we distort space to greater degrees, time compresses for them as well. Taking this into account, it thoroughly explains the ability of the craft to move at enormous speeds without signature sonic booms or exhaust trails. It also eliminates the harmful effects of inertia upon the craft and those within. The Sport Model would be capable of instantaneous acceleration/deceleration and could easily appear to make turns at right angles.

"Imagining that you're in a spacecraft that can exert a tremendous gravitational field by itself, you could sit in any particular place, turn on the gravity generator, and actually warp space and time and 'fold' it. By shutting that off, you'd click back, and you'd be at a tremendous distance from where you were, but time wouldn't have even moved, because you essentially shut it off."[4]

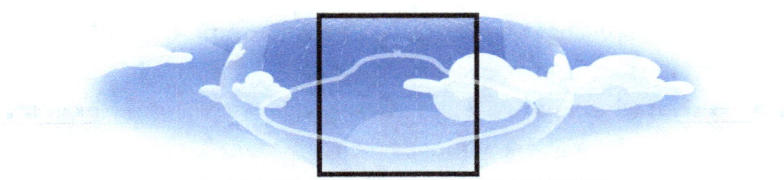

LIGHT IS BENT AROUND THE SPACE/TIME DISTORTION

LOW OBSERVABILITY

With the craft's warp bubble at "maximum distortion" some wavelengths can be directed around the craft. That means everything from radar to visible light can be rendered invisible. Observers and instruments would only see what was behind the distortion. This can be dialed up or down.

RICCARDO STORTI'S CORROBORATION

In an eye-opening October 2022, Tim Ventura podcast, Riccardo Storti (B.Eng) examined three of Bob Lazar's claims, which Riccardo states anyone could verify using mathematics. Riccardo has experience working within the research and development department of Delta Group Engineering. He has published many papers with ResearchGate.net.

Riccardo's equations show 99.9% of Earth's gravity is negated within the first 415 harmonic modes in a quantized Fourier spectrum describing a gravitational acceleration field (see *Advance Understanding* later in this chapter).

- **Harmonic Mode:** a specific pattern or oscillation that is a multiple of a fundamental frequency.
- **Quantized Fourier Spectrum:** a mathematical representation of a signal or function that breaks its factors into its constituent frequencies, represented as a graph or plot. When "quantized," it means the frequencies are restricted to certain discrete values or "quantum states." This is like how energy levels in atoms are quantized, meaning that only certain specific energies are allowed.
- **Acceleration Field:** a way to describe how the acceleration of an object changes as it moves through space, or how different parts of an object experience different amounts of acceleration.

By simulating through his mathematics, Riccardo shows how both gravity A-waves and B-waves interact. The mass of your planet or star determines how much power and at what frequency you need in order to achieve buoyancy (nullified gravity).

CLAIM 1: GRAVITY A-WAVE PULSED AT 7.46 HZ

Riccardo's mathematics determined our Earth's B-wave frequency at sea level to be 7.43 Hz. Now, Bob's frequency of 7.46 Hz is just shy of that ideal range and achieves 99.917% negation; whereas, Riccardo's ideal frequency of 7.43 Hz achieves 99.9608%. What's significant between the two rates is how achieving those last hundredths of a Hertz requires a thirty-sixfold power increase. That's a ridiculous sum that achieves very diminishing returns.

Gravity B-wave

Gravitational Environment	Pulse Frequency (Hz)	Amplification Factor*	Power Amplification Factor**
The Moon (Surface)	9.712	7.750×10^3	3.921×10^6
The Earth (Surface)	7.433	4.679×10^4	1.812×10^6
Jupiter (Cloud Surface)	2.022	1.184×10^5	1.247×10^6
The Sun (Surface)	0.989	1.309×10^6	6.744×10^6
Supermassive Black Hole***	1.249×10^6	7.266×10^7	3.098×10^{38}

* Increase in the amplitude of the gravity A-wave within the resonator(s) prior to release via waveguide(s) ** Increase in the power of the gravity A-wave within the resonator(s) prior to release via waveguide(s) *** 0.25(mm) outside the event horizon of a Supermassive Schwarzschild Black Hole (10^9 solar masses, non-rotating, non-charged)

Gravity A-wave

Harmonic	(%) Gravity Nullification	Pulse Frequency (Hz)	Amplification Factor*	Power Amplification Factor**
N = 15	97.470	0.267	1.148×10^5	3.093
N = 215	99.813	3.851	8.991×10^4	1.305×10^5
N = 415	99.917	7.433	4.679×10^4	1.812×10^6
N = 615	99.930	11.015	3.162×10^4	8.739×10^6
N = 815	99.946	14.598	2.388×10^4	2.695×10^7
N = 1015	99.961	18.180	1.918×10^4	6.484×10^7

NOTE: The highest possible Harmonic Mode is $N = 1.450 \times 10^{28}$ * Increase in the amplitude of the gravity A-wave within the resonator(s) prior to release via waveguide(s) ** Increase in the power of the gravity A-wave within the resonator(s) prior to release via waveguide(s)

If Bob Lazar was looking to fake his way through fantastical scientific claims, the first instinct would be to give only right answers. But so often in the real world, the "right" answer isn't a realistic one. Practicality steers an engineer towards a more efficient middle space. If it costs $10 for a jar of 99.92% pure honey, would it be worth your while to pay $360 for a bottle that was 99.96% pure? Your taste buds wouldn't distinguish enough of a difference for the cost/benefit to ever be worthwhile.

"To me, I think you've got a higher chance of winning the lottery than fluking that number of 7.46, to use Bob's figures. You also want to consider that the harmonic limit in this spectrum is 10^{28}. So, there's a huge number of modes, and you only need the sum of the first 415 in order to reproduce gravitational acceleration by greater than 99.9 percent."[5]

In a nutshell, if Bob was attempting fraud, he'd likely say the craft emits at 7.43 Hz. However, Bob asserts he doesn't understand why the Sports Model uses 7.46 Hz. Despite his ignorance, it turns out to be a more efficient use of energy.

CLAIM 2: GRAVITY WAVE AMPLIFIERS OPERATE IN A MICROWAVE FREQUENCY RANGE

Early on, Bob was tight-lipped about what frequency element 115 resonated at. Yet, he shared that the A-wave was accessible because its influence extended an inch from the nucleus. While it's simple enough to deduce the wavelength's measurement, Riccardo confirmed several more aspects. Working with Fourier spectrum analysis, he found an 11.44 GHz microwave frequency checked several boxes.

An additional piece of support to Bob's claim is how he described only an infinitesimal amount of that wave as usable. A factor of Riccardo's analysis shows the vast majority of harmonic modes (99.9985%) do not project beyond the nucleus, thus confirming what's accessible for amplification is only that last .0015%.

CLAIM 3: ELEMENT 115'S MACHINING PROCESS

Bob admits to seeing "no rhyme or reason" to element 115's milling process. Milling is a form of material removal by which a cutting tool, typically a rotating cylindrical cutter, is used to shave materials into a desired shape.

The S-4 facility's source of 115 was provided to Los Alamos Labs as a series of discs. It turns out Riccardo did manage to find a few rhymes and reasons.

Element 115's A-wave appears to natively phase shift as it extends away from the atom. This makes handling a purified quantity problematic because of its repulsive force.

Since the element resonates on a microwave frequency, a separation into thin discs a quarter of its wavelength effectively shuts off the gravity A-wave on its face. That makes transport simpler, without having to contend with most of its effects. A similar function happens with the window of your microwave oven. Given the oven's wavelength, the holes in the glass' mesh are too small for the waves to fit through. This allows you to see inside the oven without receiving an eyeful of microwaves.

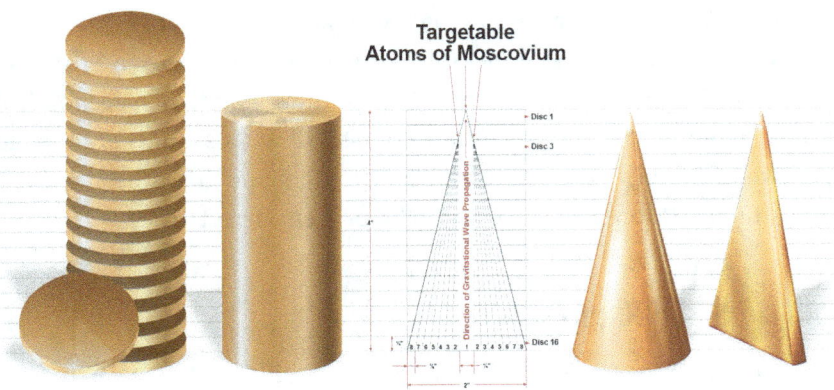

The lab could have used a couple of methods for merging the discs. Either they achieve their melting point (1,740° Celsius), or possibly pressure-fit them to forge sixteen discs into a cylinder. Next, they milled the cylinder into a conical shape. The last-stage slices a series of one-eighth wavelength partitions, leaving a primary wedge that's a quarter of wavelength thick. The remaining one-eighth slices can be paired later to form ancillary wedges of an equal quarter wavelength width.

All wedges drive the A-wave to resonate toward the wedge's tip. With this method, our reactor has an optimized point where single atoms of 115 can be targeted by the accelerator.

Remember earlier how lucky one must be in order to guess a data point like 7.46 Hz? Consider this: It's one thing to get an astronomical hunch when you have a vague understanding; it's quite another to provide data on a subject you do not understand (rhyme or reason) and still have those facts turn out to be proven correct.

WHY IS THE REACTOR REPULSIVE?

A curiosity arises when Bob describes his inability to touch the reactor when it is activated. It emits a strong enough A-wave from its hemisphere to be considered a force field. No matter how hard he tried to push against it, Bob couldn't get closer than twenty inches (50.8 cm) to the surface. However, this effect was happening before the A-wave had been routed through the gravity amplifiers.

Best guess? The reactor is configured with a meta-material that resonates and enhances its native repulsive effect.

ADVANCED UNDERSTANDING: GRAVITY WAVES

The provided descriptions in this chapter depict gravity waves in a solitary fashion for ease of understanding. However, there are no single gravity waves emitted from the Earth or by atomic elements. Here, the rabbit hole delves down into the quantum. Riccardo shares his deeper understanding by visualizing as follows:

- The amplitude of a gravity sinewave is a visual representation of the probability of virtual particles emerging from the quantum vacuum into physical reality; the greater the amplitude, the higher the probability.
- E-115's amplified gravity A-wave [at 11.44 GHz being pulsed at 7.46 Hz] is a Quantum Vacuum Inhibitor (QVI).
- This QVI blocks virtual particles of 7.46 (Hz) from popping into existence. When employed at the Earth's surface, it nullifies 99.92% of gravity.
- E-115 switches off the Casimir effect. Without the Casimir effect, a gradient cannot form in the spacetime manifold. Therefore, spacetime curvature is flat, and nothing falls.

SUMMATION

Even if we don't have the advanced mathematics and understanding of Fourier equations, Riccardo's explanations of several of Lazar's missing puzzle pieces goes a long way toward cementing probabilities.

It also shows a key complaint that Bob has about the high secrecy encasing the government's back engineering attempts. Real science requires an open engagement of ideas, hypotheses to be challenged, and experiments to be replicated. Scientists must be allowed to have dialogue with others. No two isolated physicists are going to have all the answers. You need the experience and knowledge of many field experts to complete the understanding. Otherwise, you have the universe's most prized puzzle box with no idea how many pieces are missing and where to look for them.

Bob Lazar's Sport Model accounts for all Five Observables. It is the most interesting, detailed, and studied craft by far. Hence, why it is our first case.

CHAPTER 6
T. TOWNSEND BROWN'S GRAVITATOR

BORN IN 1905 IN ZANESVILLE, Ohio. Thomas was blessed with an affluent family. His parents not only inspired his early childhood but also provided the means to build his own lab. Similar to other prodigies of his era, Thomas excelled at Science and History but underperformed in other studies.

By the age of 12, he had built his first wireless telegraph.

While attending Denison University, Brown came under the mentorship of physicist and astronomer Paul Biefeld. This professor not only laid a strong foundation for experimental techniques but also encouraged Brown's unconventional ideas. This led to his honorific naming in T. Townsend's seminal work, the Biefeld-Brown Effect (BBE).

We'll get to that in our theory section soon enough.

Now, if William Stephenson could be considered a real-life inspiration for Ian Fleming's "James Bond," then T. Townsend Brown would be an equivalent take on "Q." Here's why.

Townsend became a scientist deeply entrenched in covert intelligence. During his second active duty stint in the Navy, he was brought onboard with the Glenn Martin Co. (today's Lockheed Martin) as a materials and processing engineer. Then, toward the end of WWII, Townsend participated in a behind-enemy-lines covert mission to recover Nazi technology and scientists.

After returning to the States, Brown launched a classified project in the 1950s, codenamed Winterhaven. His primary drive was the realization of his BBE theory into actual non-conventional aircraft. That research flew well under the radar by employing a combination of private investors alongside other government sources.

In 1956, Townsend founded NICAP (the National Investigations Committee On Aerial Phenomena), which he co-chaired with Major Donald E. Keyhoe. This was a scientifically oriented organization dedicated to investigating UFOs. While Brown provided the initial scientific impetus, Keyhoe became its most prominent leader and advocate.

Major Donald E. Keyhoe

Early into the 60s, Project Winterhaven ended, seemingly with no technological breakthroughs. T. Townsend Brown's theories were discredited as the result of ionic wind. Brown's own fluid dielectric 'lifters' were cited as proof of this. However, this device became what he called his "Wounded Prairie Chicken Routine." Brown would intensionally present evidence which would discredit himself. *Why, you ask?* This way, real discoveries could go 'black' and unnoticed.

Thus, he moved on to other projects.

Before ending Winterhaven, Townsend frequently met with several prominent figures in Nassau, Bahamas. Eldridge Reeves Johnson, the predecessor of RCA Records, hosted these get-togethers. Those believed to be in attendance included Agnew Bahnson, Laurence Rockefeller, Howard Hughes, Henry Luce, and Vannevar Bush, among others. Their consortium of fellows came to be covertly referred to as the Caroline Group, named after Johnson's yacht.

Any technologies that might have come out of Project Winterhaven would have been quietly shepherded over into private industry and further developed for military and aerospace purposes.

Despite the Biefeld-Brown Effect being soundly dismissed, Townsend was still privately giving demonstrations in 1967. This time to General Curtis LeMay, Chuck Yeager, and Edward Teller (father of the hydrogen bomb). Townsend's daughter Linda related how Teller was dumbfounded by her father's gravitators. He even stated, "I have no idea how this works." At which point Teller's wife responded, "You don't know how nice it is to hear him say that."

T. Townsend Brown died in 1985. A few years later, his wounded prairie device (the fluid dielectric lifter) was licensed to the Sharper Image company. It was then modified and telemarketed as the Ionic Breeze.

NOW, HERE'S THE TWIST

You might wonder, what could inspire a youth to grab hold of such an innovative notion as controlling gravity and doggedly pursue it for his entire life? Well, after Brown's passing, his daughter confided in what he shared with her.

While living in California, between 1923-24, a teenage Thomas had a UFO encounter on a Catalina Island ridge. An orb appeared in the sky and approached him.

"He learned so much standing there with that ball of light that he went back to his (sic) lab in Pasadena. He went to work immediately. That was the beginning of his life's work. And he said that everything he ever learned about his work he learned instantly." —Linda Brown-Towne

THE BIEFELD-BROWN EFFECT

This theory is at the root of all of Townsend's efforts. He believed there is an underlying connection between electro-magnetics (EM) and gravity (G). If given the right combination of EM forces, one could affect the other. Toward that end, he built an experiment to support his hypothesis.

THE GRAVITATOR:

1. Create two metallic discs (solid dielectric capacitors)
    a. one smaller than the other
    b. separate with a neutral insulator rod
2. Run electricity through both at:
    a. extremely high voltage (1 million V)
    b. but extremely low current (0.00001 to 0.00005 Amps)
3. Small disk is charge: positive (+)

4. Large disk is charge: negative (-)

These conditions result in the larger plate becoming attracted to the smaller plate while the smaller plate is repelled.

Hold up! That is not how opposite charges normally react.

When you take two positively charged magnets and bring them together, they inherently push apart. Now, imagine you got the same effect when you sat a positive and negative magnet close together. The negative magnet is naturally drawn to its positive counterpart, but the positive magnet is suddenly thrust backward. Such a result would lead us into a runaway chasing effect.

As a last step, take the two disks and stack them smaller (+) over larger (-). *Then, yep. You guessed it.* The chase overpowers gravity, causing them to rise.

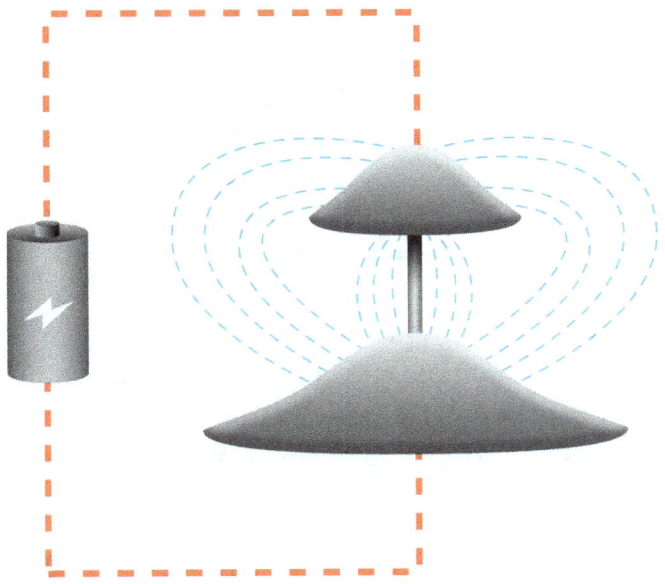

Detractors of this theory emphatically declare it is because of ion wind. That is the flow of air generated by ion movement between two electrodes. This is caused by collisions between accelerated ions and neutral air molecules. However, this also depends on the presence of air molecules for ion collisions. Ion wind is not a factor in outer space travel. It cannot occur in a vacuum.

Brown attests to performing partial vacuum tests in his career. NASA has even stated their attempts to recreate his theory in a vacuum failed to confirm Brown's results. But consider that NASA is an adjunct to the Dept. of Defense. Overlapping motives might lead one to believe if BBE had been classified, this would make an ideal way of dismissing it. As of yet, no independent vacuum tests have confirmed Brown's results. Podcaster Jesse Michels has even offered a $50,000 reward for verifiable or falsifiable proof. Yet, that might not be enough of an enticement.

Performing Townsend Brown's Gravitator test in a vacuum is challenging for several reasons due to technical, financial, and experimental design limitations. Here's why:

- **Cost and Equipment:** To determine if the effect is because of ion wind or something else, the vacuum must be strong enough (high vacuum, below 10^{-5} Torr) to remove nearly all air molecules. Large, high-vacuum chambers can cost tens to hundreds of thousands of dollars.
- **High-Voltage Issues in a Vacuum:** In a low vacuum (partial pressure), high-voltage setups may cause uncontrolled dielectric breakdown.
- **Power supply:** Generating 1 million volts in a vacuum without arcing is difficult and requires specialized high-voltage power supplies.
- **Insulation Challenges:** Many materials that work in air fail in a vacuum due to charge buildup and the lack of proper insulation.
- **Instrumentation:** Measuring tiny forces (if present) requires ultra-sensitive torsion balances or laser interferometers, adding more cost

However, in 2003, the Army Research Lab[1] performed experiments on four different-sized asymmetric capacitors. Those results showed the ionic wind was at least three orders of magnitude too weak to account for the observed force. In other words, the force moving their capacitor was only 1 part ion wind and 999 parts something else — an undetermined force.

UNDERLYING PHYSICS

Townsend contended that through BBE, he had uncovered a linkage between EM and gravity. He believed that a gravitational gradient/wave was forming, powered by the secondary effects of charge flowing through his capacitors. To grasp why and how, we need to understand what the capacitors are made of.

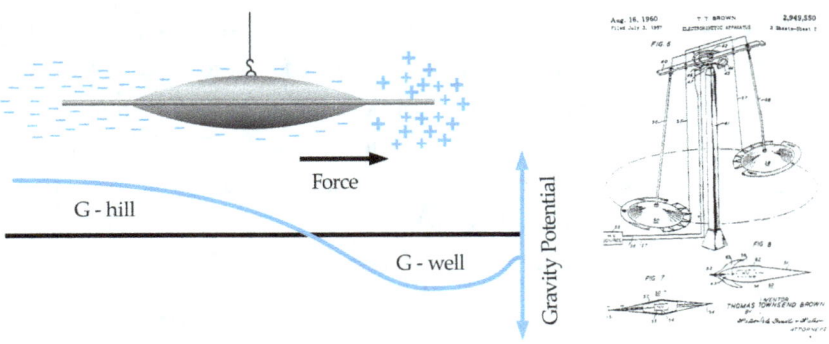

Dr. Brown was, first and foremost, a material scientist. He focused on finding elements that could create novel properties when a current was applied. For that reason, he turned to bismuth (Bi).

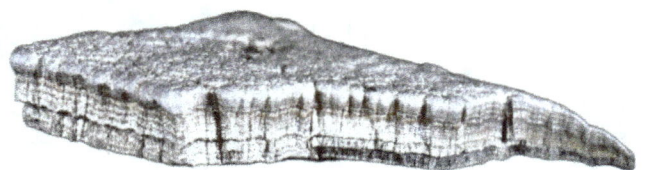

Art's Parts: Layered Bismuth Magnesium

Those in the UFO-know find that detail compelling. There have been several anomalous crash retrieval samples composed of bismuth. As an example, 'Art's Parts' was analyzed by Stanford Prof. Garry

Nolan. His scans revealed extremely thin layers of bismuth (1–4 microns thick) bonded to thicker layers of magnesium. For comparison, human hairs run between 70 to 100 microns in thickness. Such manufacturing today would cost obscene sums to replicate.

Another material Townsend experimented with was barium titanate ($BaTiO_3$). Both bismuth and barium have what's called a 'high k-factor,' which is the ability to store electrical energy. Materials with a high k-factor exhibit strong polarization, making them ideal for devices like his capacitors.

To contain such high energy, one needs to insulate them. For both materials, Townsend sealed with layers of aluminum, specifically aluminum oxide (Al_2O_3). This creates a barrier against leakage.

It is also suggested that the form of bismuth Townsend employed was not a standard one. It turns out to be a match again to Garry Nolan's scans, magnesium bismuth (Mg_3Bi_2). Louis Witten, a man famous for theoretical physics in general relativity and gravity, also confirmed observations that Townsend had found a type of bismuth that was repelling instead of attracting.

But what does Witten mean by repelling?

This can get very confusing. *I'll do my best here.* Most readers are probably familiar with positive and negative charges stemming from electricity and atoms. However, when it comes to the overall concept of matter, all matter is considered to be mass-positive. Don't think of it as a charge, instead an addi-

tive feature of all material. The more mass you add, the more its positivity strengthens. Thus, all mass is attracted to itself through gravity.

In a 1956 Chapel Hill conference, Austrian mathematician Hermann Bondi released a paper titled *The Role of Gravitation and Physics.* In it, he explored the theoretical implications of negative mass within the framework of Einstein's General Theory of Relativity. His equations revealed that in a system containing both positive and negative masses, they would exhibit this same runaway motion, where the negative mass chases the positive mass, and the positive mass flees from the negative. That perfectly matches Brown's asymmetric capacitor experiment.

Why has this never been determined before?

EXTENDED ELECTRODYNAMICS (EED)

Of the four known forces in the universe (electromagnetics, strong force, weak force, and gravity), we only have control over one: electricity. And that control stems from the equations of James Clerk Maxwell, Oliver Heaviside, and Edward Lorenz. These physicists/mathematicians worked out the formulas we use today as classical Electro-Dynamics.

However, to get their math to work, they had to simplify and assume certain figures. Maxwell's original 20 computations included a **scalar field**. Remember the **Higgs field** from our Starting Principles chapter 3? **That is a scalar field.** Think of it like an invisible plot of numbers assigned to every point in an infinite-sized cube.

Formulation of Maxwell's equations in terms of vector and scalar potentials

$$\nabla \cdot \mathbf{B} = 0 \qquad \Rightarrow \mathbf{B} = \nabla \times \mathbf{A}$$

$$\nabla \times \mathbf{E} + \frac{\partial \mathbf{B}}{\partial t} = 0 \Rightarrow \nabla \times \left(\mathbf{E} + \frac{\partial \mathbf{A}}{\partial t} \right) = 0$$

$$\mathbf{E} + \frac{\partial \mathbf{A}}{\partial t} = -\nabla \Phi$$

$$\text{or } \mathbf{E} = -\nabla \Phi - \frac{\partial \mathbf{A}}{\partial t}$$

The importance for EED is that Maxwell's scalar field is not apparent when brought over into Heaviside's simplified calculus. The next scientist, Lorenz, added a rule that made it even harder to see this hidden function. All of this was for simplicity's sake, making it easier for other scientists to understand the electromagnetic force.

When we bring back the scalar field, we discover three new kinds of waves with wild potentials. These differ in behavior from EM waves and may couple more tightly to gravity.

- **Scalar:** Think of ripples in a pond, but instead of moving up and down or side to side, they just expand and contract in place. Unlike regular EM waves (like light or radio waves), scalar waves don't lose energy as they travel.
- **Scalar-longitudinal:** Waves that push and pull in the same direction of motion. Think of this like a rope, where a tug on one end is felt on the other, regardless of distance (instantaneously).
- **Helicoidal:** Waves that twist as they move, like a corkscrew or a spiral staircase. They combine both twisting and forward motion and again without energy loss as they travel.

5 DIMENSIONS

With General Relativity, the consensus has been we live in 4-D space-time, where three dimensions are spatial (height, depth, width) and time is our fourth. String theorists seeking to unify Relativity and Quantum Mechanics currently must resort to creating an extra 6–7 hidden dimensions for their mathematics to work. Today, daring physicists are discovering that EED may lead to similar results but only adds one hidden spatial dimension.

5 Dimensions

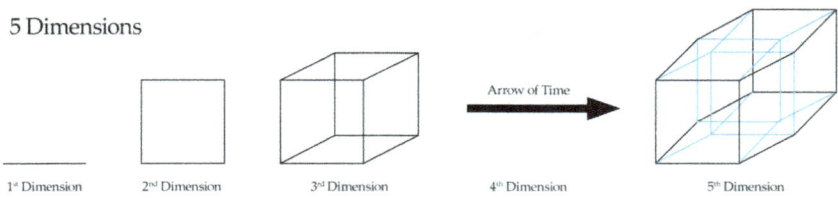

1ˢᵗ Dimension   2ⁿᵈ Dimension   3ʳᵈ Dimension   4ᵗʰ Dimension   5ᵗʰ Dimension

In essence, this fifth dimension provides an expanded mathematical framework where the energy and momentum of EM vortexes can couple more directly with spacetime curvature, thereby influencing gravitational behavior (G). Without the fifth dimension, this coupling is either impossible or highly suppressed.

These factors are why the Biefeld-Brown Effect and Extended Electrodynamics are so powerful, perhaps even too powerful. The physicists who have rediscovered its nature warn that unimaginable weapons inevitably spring from this understanding. Thus, an equally powerful reason to classify this field of Physics.

Lodge this chapter in the back of your mind as you continue forward with the chapters ahead. It will be hard to miss its continuing thread of influence.

CHAPTER 7
BACKSTORY: MARK MCCANDLISH

THE YEAR IS 1988, and aeronautical illustrator Mark McCandlish is enjoying industry success. His contracts with Lockheed Martin, Northrop, McDonnell Douglas, General Dynamics, and more not only provide him classified access, but have kept him so busy illustrating he doesn't have the capacity to attend a restricted, Norton Air Force Base (AFB) air show.

But his associate, Brad Sorenson, could go in his place.

Upon Brad's return, he seemed quiet; a bit shaken. Gradually, he opened up about inadvertently wandering into something he wasn't supposed to see.

After falling behind his tour group and catching up with another, Brad's new group was shuttled to a hanger. An Air Force general's presentation ended with the revealing of three hovering disc-shaped craft of varying diameters: Baby Bear (24 ft; 7.3 m) diameter; Momma Bear (60 ft; 18 m) diameter; and Daddy Bear (130 ft; 40 m) diameter. For simplicity's sake, we'll focus on the smallest of the three. Commonly referred to as an Alien Reproduction Vehicle (ARV), Lockheed Martin dubbed it the *Flux Liner*. It is through the manipulation of the quantum vacuum that this craft would not only cancel out the its own mass, but as a result would allow faster-than-light travel.

Brad roughed out a loose sketch, which Mark reworked into a detailed schematic. As fascinating as this was, Brad became more unnerved by the possibilities. He considered Mark's blueprint "dangerous."

While the thought-experiment thrilled Mark, he understood it was the testimony of just one person. That was until the early 1990s, when he ran into Kent Sellen at another air show.

Kent related, back in 1973, he had been an air chief at Edwards Air Force Base. On his way to a job site, Kent passed a hanger with its door slightly open. He peeked in and was shocked to see a hovering flying saucer. Before he could gather his wits, he had a gun to his face and a bag over his head. He spent the next eighteen hours being interrogated, before being released.

Without prompting from Mark, Kent's description of the craft matched what Brad had shared. Grabbing a lens-cleaner pack, Mark sketched an outline. Kent's response: "Oh! You've seen it too."

At least one other person was willing to corroborate Brad's assertion about this craft. Lt. Col. John Williams was aware of a facility at Norton Air Force Base that was considered a no-go zone. The standing rumor was that this was a storage facility for a UFO.

Later, Mark met up with Lt. Col. Wendelle Stevens, a USAF pilot that chased craft over the arctic. When asked if Wendelle had ever taken a photo of one of these chases, he presented this pic from 1967.

Over the decades since, Mark has encountered several people that have shared bits and pieces. These tidbits eventually helped him perfect his designs into what we'll share in the next chapter.

Unfortunately, in 2021, Mark McCandlish was found dead from a self-inflicted 9mm gunshot (not shotgun). I mention this to dispel rumors of Mark being 'suicided.' Sadly, the man suffered from several physical ailments and had been medicated for depression. The COVID-

19 pandemic seemed to have worsened his condition. He had confided in his brother the week prior to ending his life.

All evidence points to a very real choice on his part.

CHAPTER 8
MARK MCCANDLISH'S FLUX LINER

THIS CRAFT WAS DESIGNED to harness advanced electromagnetic and gravitational principles to achieve propulsion and maneuverability. The term "flux" refers to the flow or movement of something, particularly in the context of magnetic or electric fields. McCandlish's understanding was that the Flux Liner utilized controlled fluxes of electromagnetic and gravitational fields to counteract the Higgs field and enable the craft to travel through space.

We're talking about Lockheed Martin's first-generation prototypes, which, by the late eighties, had already been in use… for decades.

OVERALL APPEARANCE AND LAYOUT

Unlike the previous chapter's sleek Sport Model, the Flux Liner looks more like a clunky, diving-bell-style construct from the late fifties. Indeed, the surfaces appear old and weathered. The outer coating is a "goopy" lead paint, showing scratches and scuffs. The exterior dome cameras were off-the-shelf components used for closed circuit cameras in Vegas casinos. As well, the crew cabin's hatch is a classic submarine door.

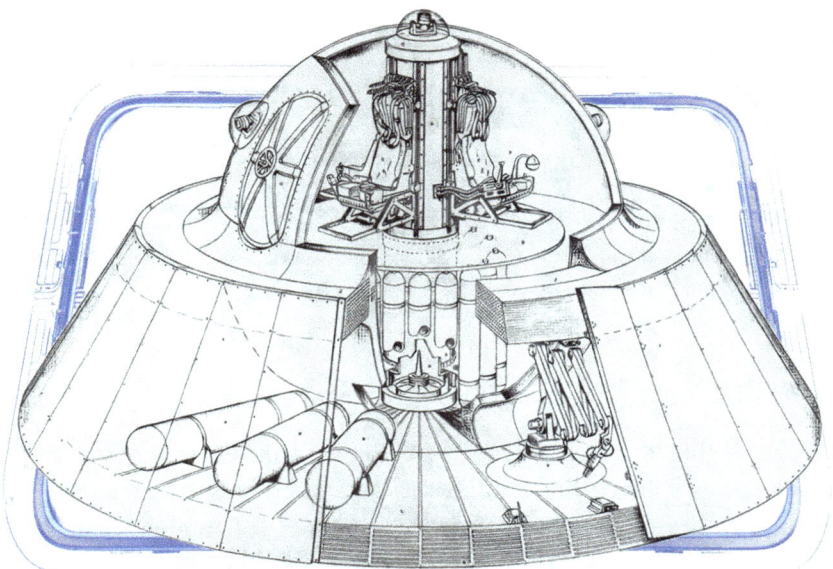

To understand the principles at play, we'll focus on the smallest of the three craft. The overall function is not dissimilar to a Tesla coil. In those terms, it comprises a feedback loop between a capacitor (battery) and a resonant transformer/antenna (magnetic coil). The major difference is the amplifier and flywheel running between them. The electromagnetic interchange between these four devices produces a force which not only alters the craft's mass, but creates a local spacetime bubble.

What follows is based upon Mark McCandlish's presentations and email exchanges with Paul A. LaViolette, PhD.

ENGINE STRUCTURE

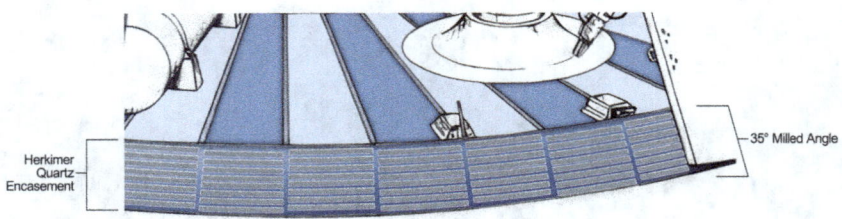

Herkimer
Quartz
Encasement

35° Milled Angle

CAPACITOR

Seated within the base of the craft is our capacitor. There are forty-eight arrays splayed out around the center, laterally staggered beside their neighbor. Each triangular array is made up of an eight-plate stack of conductive metal. This is composed of a foil lamination made of a Dow Chemical alloy (AZ31-X) that is 95% magnesium and 5% zinc, bonded with bismuth, then skinned in a thin copper cladding.

The end plates are milled at a 35° angle, becoming progressively smaller, one over the other. Our capacitor is then encased in dielectric Herkimer quartz. This form of quartz is normally clear, but exposure to high radiation has led to a green staining, as noted in one schematic.

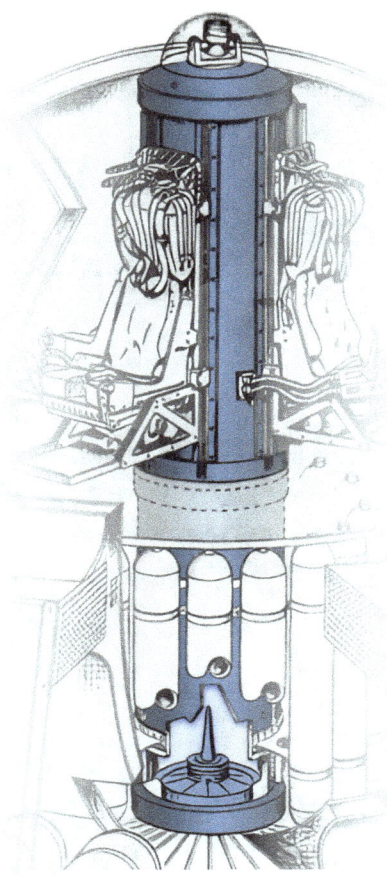

AMPLIFIER

What extends up from the capacitor through the crew compartment is considered the most secretive component. Over Mark's delving and parsing details, blanks eventually filled in.

It comprises two quartz cylinders (an annular duct) inlaid with copper filaments that spool from peak to base. Flowing between these cylinders, mercury, as a pseudo-noble gas or plasma, is circulated in a closed loop by the capacitor's applied electricity. The mercury descends within the inner cylinder, then back up in the space between cylinders. They sheathed this within an outer layer of secondary copper windings. Both internal cylinders and mercury are locked in counter rotations of each other.

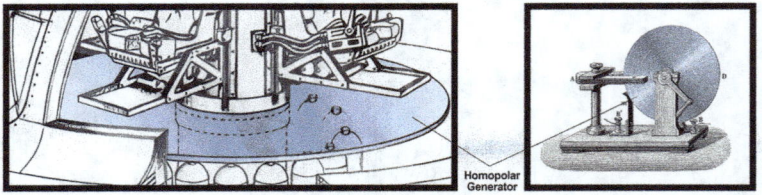

FLYWHEEL

Riding midway up the amplifier there appears, at first glace, to be a nine-foot (2.7 m) diameter flywheel. This is a homopolar generator, one of Michael Faraday's inventions. The centrifugal force of its spin within a magnetic field drives electrons toward the wheel's edge. A conductive brush then taps this surplus power for use.

ANTENNA

Around the beltline of the crew compartment is a two-foot (0.6 m) thick, dielectric Herkimer quartz encasement of coiled copper wire (resonant transformer). This converts generated electricity into magnetic waves. From here, an extremely high-voltage, magnetic toroidal field loops around through the capacitor and back into the amplifier.

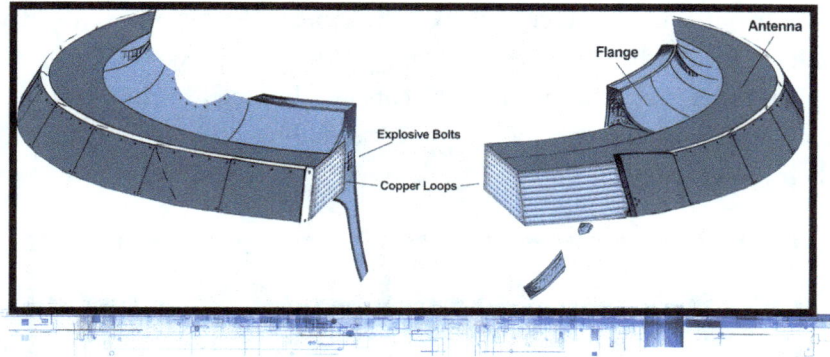

Our antenna is seated into a flange, retaining it to the crew compartment. A series of explosive bolts are embedded in the flanges that can sever the link in an emergency.

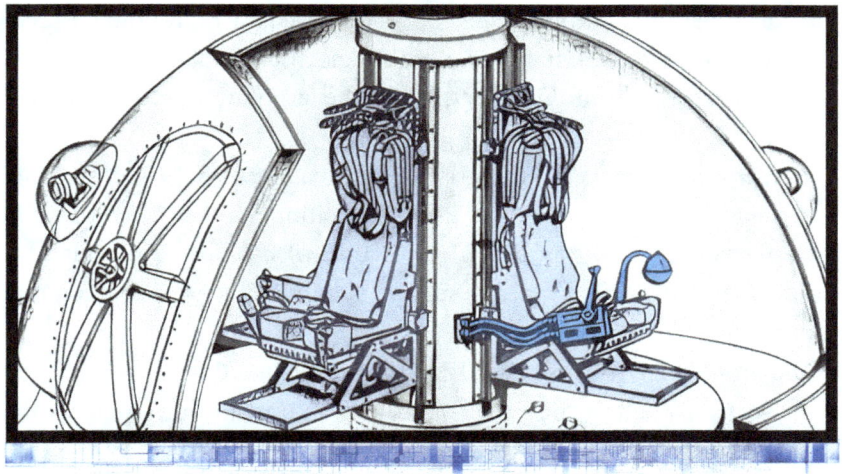

CREW COMPARTMENT

Ejection seats are rail mounted back-to-back, around the amplifier's outer casing. Built into the seat base is a pan for crewmembers to rest their feet, so as to not contact the spinning flywheel beneath.

In the event of an emergency ejection, the antenna's explosive bolts separate the compartment from the disc, pulling the amplifier's outer sheath out with the crew sphere. The sphere doubles as a re-entry vehicle for Earth's atmosphere. Once below 15,000 feet (4572 m), the sphere pops a chute. The ejection seats are designed to ride down on their rails, out through the bottom. Free-falling crew seats then pop their individual chutes.

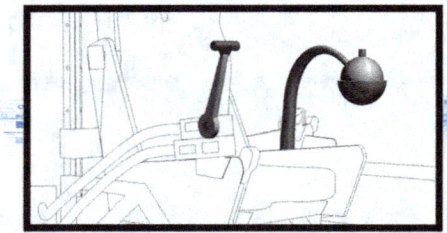 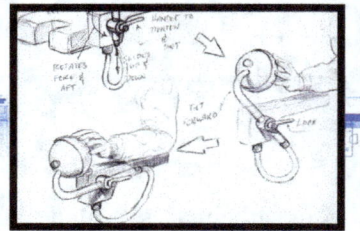

FLIGHT CONTROLS

The pilot has two primary controls. The first is a high-voltage lever (potentiometer, or rheostat). Heavy-duty electrical cables run from this to the central column. This regulates the amount of power in the system.

On the opposite side, there is the equivalent of a 1950s trackball. The pilot has an adjustable hook-arm to position it either forward or on an up position and then clasp-lock it into place. The arm ends at a capped sphere. Riding on the sphere is a bowl, which acts a pivot control.

Fiberoptic leads (first publicized by Corning Glass in 1970) are inserted from the arm into the sphere. These transmit inputs from the sphere's inner sensors to the forty-eight capacitor arrays. Within the bowl's center is a laser diode that scans the sensors to know which capacitor plates to strengthen. This opens or closes a series of relays, which streams electricity throughout the capacitor.

Do note, only one set of pilot controls are visible in the schematic. Perhaps a redundant set could be on the alternate side. As for the remaining seats, nothing was mentioned about that crew's function.

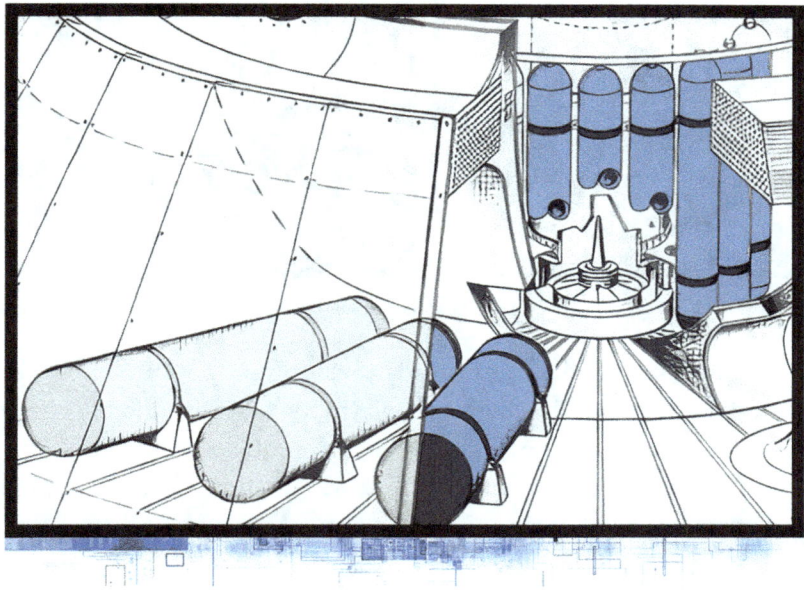

AIR SUPPLY

Strapped vertically around the amplifier's lower half, below the flywheel, are twenty-four atmospheric tanks. These are connected to even larger tanks atop the capacitor. All these tanks are protruded, composite vessels, devoid of steel or aluminum. The crew seats also have small tanks under each, in case of ejection.

SYNTHETIC VISION SYSTEM

Seven pressurized, exterior, glass bubbles (one on top and six surrounding) contain off-the-shelf surveillance CCTV cameras. Two cameras at a time feed video to our crew through a headset. The separate feeds not only create depth perception, but can toggle views based on head position.

Concept mock-ups of cameras and VR headset

Okay. Cool. But why this instead of windows?

Turns out the extreme high voltages in the outer field ionize the air. This ionization is so strong, x-ray photons produced would be at lethal levels for the craft's occupants. So a dense radiation barrier is required to protect them.

DOOR

This is your classic, steel, submarine door. A wheel lock controls its pins into the door frame. The retaining frame creates an airtight seal. Likely, additional dense materials were layered within the door to act as radiation shielding, since steel does not effectively protect against x-rays.

ROBOTIC ARM

A pneumatic (air) claw sitting atop the capacitor floor allows for exterior retrievals. You may wonder, why air-driven? With the electromagnetic field so intense, conventional electric motors will not function. So, our crew resorts to pistons, gears, and air. It's great for when you want to return with lunar or Martian samples.

OUTER PANELS

The panels are a fiberglass composite material. These are connected with Dzus fasteners, allowing for quick affixing and releasing. As added protection, lead-flecked paint slathers every surface, save for the door and camera bubbles.

SO, HOW DOES IT WORK?

The main problem with attempting to recreate an alien craft like Lazar's Sport Model is that our current material sciences are incapable of producing metamaterials on such an atomic scale. But that doesn't mean new physics can't still be unlocked.

As an alternative to the Sport Model, the Flux Liner's engine not only produces a spacetime bubble and electric power for its devices, this craft claims to achieve faster-than-light travel through a negation of mass.

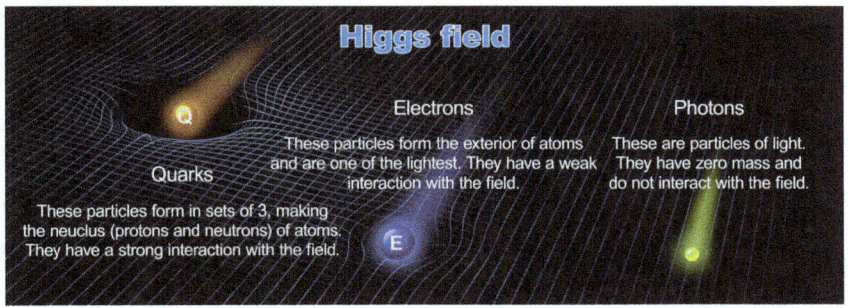

You may recall in our Starting Principles chapter that mass is assigned to matter by passing through the Higgs field. The faster you accelerate, the more your mass increases. But what if you could control your mass? What if we could dial down our interaction with the Higgs field?

This is at the heart of the Flux Liner's capabilities. After reviewing multiple sources, an understanding of how it achieves this remains not only vague but varied depending upon the source. What follows is this author's best (non-PhD) explanation.

MASS REDUCTION

Tesla's original coil patent was a step up a ladder he'd hoped would lead to unlimited free power. Stored electricity in the capacitor would form a constant loop with the resonant transformer (our antenna). The flaw was in its conductance from one to the other. Basic wires diminish the charge because of resistance/friction. This bleeds off electrons, eventually leading to a loss of power.

In the Flux Liner design, we replace a conventional wire with our amplifier. The capacitor's charge (about a million volts) is fed into mercury as a plasma, which acts as a superconductor. This means little if any loss of power as the electricity is cycled between the capacitor and the transformer.

Cylinder (annular duct) concept

But the charged mercury also generates an intense magnetic field. The counter rotating motion of the plasma is enhanced by the spinning copper-coiled quartz cylinders within.

Next, our flywheel spins counter to the amplifier. Within this intense magnetic field, it generates additional power through electromagnetic induction. (It's also called a Homopolar Motor or a Faraday disk generator.)

From there, electricity is added and poured into our quartz-encased antenna (Tesla coil). Now you have two intense electromagnetic fields. The frequency at which the two fields pulse amplifies (redoubles) the inner one upon the outer one, generating a massive, toroidal electromagnetic field, which loops around the entire craft, back though the amplifier.

It is a feedback loop that compounds power upon power.

Now, while that is the overall concept, there are vagaries surrounding how this interacts with a polarized quantum vacuum. Some imply the magnetic field creates an imbalance in the zero-point energy that can be extracted and used. However, the primary function is as a Higgs field inhibitor.

In a yet-to-be-disclosed method, high-voltage magnetic field inter-actions are applied against virtual photons that correspond between matter and the Higgs boson.

If you can reduce mass, it may translate into: the faster you go, the more available energy you have for going faster, and the more mass is reduced. A continued acceleration factor might have an unlimited effect... even beyond light speed.

Taking the reduced mass one step further; electrons cycling through the Flux Liner's (Tesla) coils would not experience friction, thus creating a lossless energy loop.

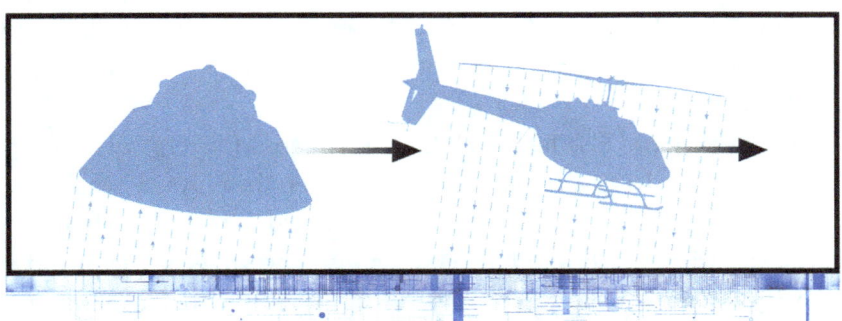

CAPACITOR AS A STEERING DRIVER

As mentioned earlier, our capacitor has relay controls for distributing electricity to various portions of the array. You can visualize its driving function as similar to a helicopter rotor.

When power is delivered equally to all the arrays and throttled higher, our craft/helicopter flies straight up. When our pilot wants to move forward, they redistributed the power balance; higher in the back, lower in the front. Just like how a helicopter's main rotor has a pivot for banking.

When power is delivered equally to all the arrays and throttled higher, our craft/helicopter flies straight up. When our pilot wants to move forward, they redistributed the power balance; higher in the back, lower in the front. Just like how a helicopter's main rotor has a pivot for banking. Rather than a mechanical pivot, we achieved this using the previous chapter's Biefeld-Brown effect[1].

SUMMATION

You can see how such a craft could be constructed based upon 1950s technology. It shows a first step into how aerospace engineers would attempt reverse engineering off-world craft.

Upon Nikola Tesla's death in 1943, at the height of WWII, the FBI ordered the Alien Property Custodian (Yes, they called it that. However, back then, "alien" had more to do with foreign adversaries) to seize his belongings and research rather than allow them to fall into enemy hands. That work remained classified until Freedom of Information Act (FoIA) requests were made between 2016 and 2018. While they released over 250 documents[2], there remains many to be disclosed.

This feels more like Tesla tech than alien reproduction, but you can see where physicists and engineers would rely heavily on what they understood and had easy access to. They would begin by studying an extraterrestrial vehicle and then relating it to known tech. The purpose of a prototype is simply trial and error; to learn from mistakes and apply them to next iterations.

Lastly, I'll plant a seed for a later chapter. Keep the Flux Liner in mind when you read Salvatore Pais' chapter. In this Flux Liner chapter, there is a quite a bit of detail on the components and interactions, but this one is light on the quantum mechanics. Dr. Pais is the polar opposite. His chapter is light on the components while detailed on the physics.

CHAPTER 9
SALVATORE PAIS' NAVY PATENTS

NEXT, we turn from the controversial to an accomplished US government-accepted source. This man is an aerospace engineer and the inventor of several patents that could be equated with UAPs.

Salvatore Cezar Pais' accomplishments in mechanical and aerospace engineering are truly remarkable. He has a PhD from Case Western Reserve University. He's worked with some of the most prestigious organizations in the aerospace industry, including NASA and Northrop Grumman Aerospace Systems.

Dr. Pais is recognized for his advanced knowledge of theory, analysis, and modern experimental and computational methods in aerodynamics. He also possesses an in-depth understanding of air-vehicle and missile design, particularly in the domain of hypersonics.

But that's not all. He also has under his belt:

- Advanced knowledge of electro-optics
- Emerging quantum technologies
- Laser power generation

- High-energy electromagnetic field generation
- Condensed matter physics

...to name just a few. It is not an exaggeration to call Dr. Pais' contributions in these fields groundbreaking. Currently, he holds a permanent civilian position with the Department of Defense, Department of the Navy, and Strategic Systems Programs (SSP).

Springboarding off of Maxwell's heavy side equations, Dr. Salvatore Pais is looking at Einstein's relativity from an atypical perspective. And through this novel lens, he believes a super force could bridge the divide between general relativity and quantum mechanics.

Examining these inventions comes with challenges. He has one vehicle, and multiple other devices, patented with the US Navy. And it's not just that we're dealing with mathematically possible technology. Yes, Dr. Pais admits, as far as he knows, no one has experimentally confirmed this patent is functional. Rather, our real challenge is that Salvatore intentionally withholds or keeps important details vague. His reasoning is out of national security. Dr. Pais is a patriot and understandably does not want antagonistic countries to gain access.

With that in mind, lets spin this up.

Concept craft based on patent design

We'll start with two craft configurations, working from identical principles. These designs employ an enormous amount of energy to generate a local void around the hull of his vessel. That null space is meant to disrupt the connection between the Higgs field and our vehicle.

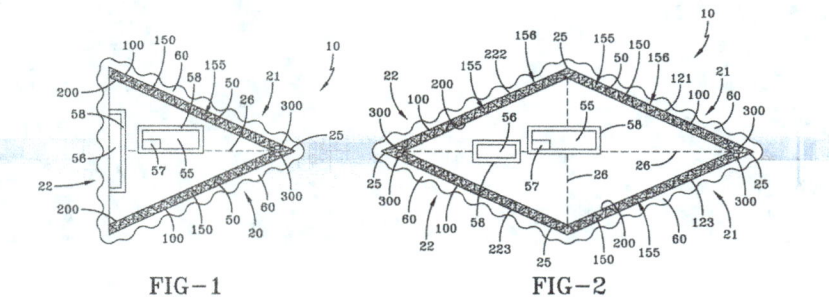

FIG-1   FIG-2

OVERALL APPEARANCE AND LAYOUT

Being a theoretical craft, this is more of a whiteboard illustration than a schematic. Beyond a pyramidal or octahedral configuration, the size and ratios have not been detailed. But Salvatore has shared his vision of a more flattened wingspan. Given the propulsion method, aerodynamics is actually less of a priority over simple landing practicalities.

SO HOW DOES IT WORK?

The intention is to reduce the craft's mass/weight, allowing it to travel at extreme speeds. The ship's combined inner (200) and outer (300) chambers generate a "high energy density" (60), which is applied against the quantum vacuum. This force would negate the craft's weight.

Whereas the Sport Model craft emphasized a gravitational wave nullification, Dr. Pais' craft is like the ARV. They both invert the focus from gravity to the mass of the ship. Essentially, all three are balancing a similar principle. It's just that their views are from different angles. Six of one versus a half-dozen of the other is effectively the same notion.

The most controversial aspect of Dr. Pais' patent is the amount of raw power his equations indicate are required to achieve this effect. The Navy denied most of his initial patent applications because, at first glance, he'd need more power than what's produced within a neutron star.

For context, neutron stars are incredibly dense and compact remnants formed after a supernova. They are composed primarily of tightly packed neutrons. If one were to magically teleport a teaspoon of a neutron star to the surface of the Earth, it would be beyond catastrophic. There'd be:

- **A massive explosion:** The sudden introduction would cause an explosion of an enormous magnitude. The energy released would be that of a fusion bomb.
- **Gravitational disruption:** The intense pull of this teaspoon would distort the Earth's gravitational field. This disruption would affect the planet's tides and seismic activity, and the orbit of the moon.
- **Structural collapse:** The immense weight and pressure of the stellar matter would cause immediate compacting of the ground beneath it. It would form a crater, accompanied by shockwaves through the planet's core.
- **Radiation:** Neutron star matter emits high-energy radiation, including X-rays and gamma rays, inducing radiation sickness in any possible survivors.

Now *that* is a lot of power. I think it helps to have some grasp of the levels we're talking about; however, it is also a brute-force or rudimentary perception of the mathematics.

According to Dr. Pais, there are four avenues one could explore to reach his high energy density: charge, voltage, spin, or vibration. The obvious two, which were dismissed, are electric charge and voltage. The third, spin, Salvatore sees as a possibility, but we run into enormous centrifugal forces.

I'll equate this to a teenage fantasy of my own.

Back in high school, I had a clever thought pop into my head. If a time traveler was inside a hoop that was spinning at near light speed, they could sit tight while time flew by around them. I imagined an intricate collection of gears, using varying sizes to spin the hoop around. In principle it was a fine thought experiment, but, in reality, there isn't a material in the universe that could hold its form under those centrifugal forces.

However, vibrations… now that's a different matter.

PIEZOELECTRICS

We use piezoelectric materials in cell phones to create vibrations (haptic feedback) when we interact with them. This technology is commonly used in touchscreens, where the user's finger pressure on the screen creates a small charge that is detected by the phone's controller. This activates the piezoelectric material to create those vibrations.

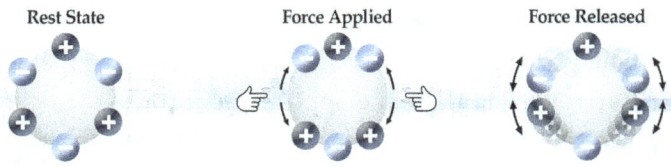

Dr. Pais' high energy density also requires his tech to use a non-linear material. What's that mean? Your basic linear material would be a metal spring. You apply ten pounds of force to the spring, and it produces ten pounds of thrust when released. A non-linear material can have a redoubling, or cascading of whatever force is applied. If you apply ten units of force, you get twenty or a hundred units in return.

While cell phone ceramics are of lead zirconate titanate, Dr. Pais' craft would rely on a different material all together. That brings us all the way back around to our familiar fourth state of matter, plasma.

ENGINE STRUCTURE

The crux of this craft is literally within its hulls. A close-up cross-section shows the components in play.

Our hull comprises an electromagnetically insulated inner wall, a charged outer wall, and a significant cavity between them. Within that cavity is a cylinder within a cylinder (annular duct). This allows for two-way flow between the inner and outer pipe. Next, we fill this cavity with our non-linear piezoelectric material. The patent suggests either xenon or another noble gas.

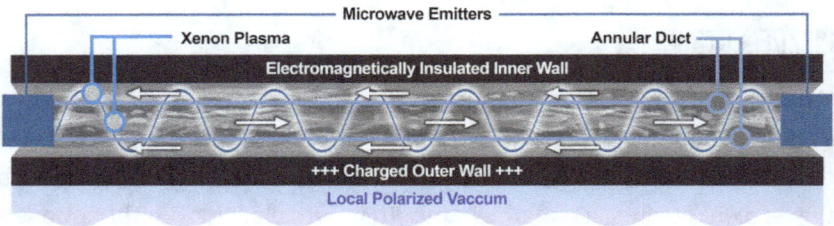

Several pivot points in the cavity are fit with microwave emitters. These fire high-frequency electromagnetic (EM) waves throughout the cavity, both heating and polarizing the material into a circulating plasma. Being polarized, the electrons orbit within one pipe, while the proton/neutron nuclei spin in the opposite direction of the other pipe.

The pulse of these EM waves accelerates, heightening the plasma vibrations far above their natural equilibrium. As more and more frequency acceleration is poured into the plasma, the maelstrom of chaos abruptly emerges into an organized structure of its constituent particles. This is called the Prigogine effect, where novel properties are exhibited differing from the element's rest state.

Here's where one of those important factors is being withheld. This plasma must not contact the cavity's encompassing walls. This cavity functions like a particle accelerator, which classically keeps protons suspended as they're accelerated in a loop trajectory. Our plasma is polarized, so my first thought in keeping it away from the hull would be similar magnetics. But since it is part of the secrecy, we don't know. In whatever method of plasma isolation is achieved, those cascading vibrations are somehow transferred outwardly.

The exterior wall may be aluminum or copper at room temperature, or a cryogenically cooled superconducting material (yttrium barium copper oxide or niobium). It is the resonant vibration in this material that achieves the needed high energy density of 8^{25} joules per meter cubed (10^{33} watts per meter square). This power generates the Schwinger effect and, thus, a polarized vacuum around the craft.

SCHWINGER EFFECT

First predicted by physicist Julian Schwinger in 1951, this phenomenon happens when the vacuum begins to boil from a strong electric field. This boiling action converts virtual particles and antiparticles into real particles, thus fulfilling $E=MC^2$. We're converting energy back into forms of matter.

The strength of the electric field required for the Schwinger effect to occur is extremely high, on the order of 10^{18} volts per meter. We have only studied such power levels with astrophysics in relation to neutron stars and black holes.

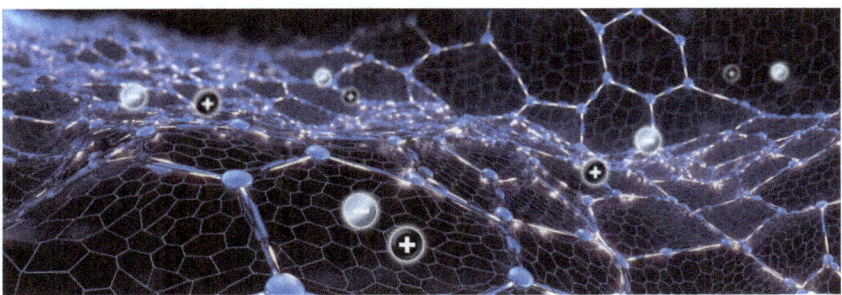

Visual representation of electron/positron materialization within a charged graphene lattice

That is until recently, when scientists applied an electric field to a lattice of graphene.[1] This incredibly strong material behaves like a two-dimensional structure. Their experiment produced, seemingly out of nothing, the equivalent of electron-positron pairs: electrons and the condensed-matter analogue of positrons, called holes. The observation matched Schwinger's predictions.

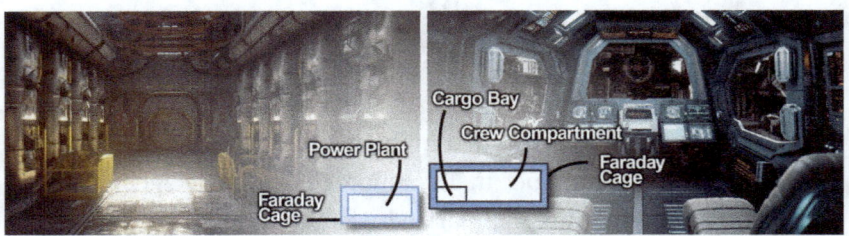

Visual mock-ups of the power plant and crew compartment

CREW COMPARTMENT AND CARGO BAY/POWER PLANT

Space is provided within the vessel for both crew and cargo. A separate section is reserved for the craft's power source. Both compartments are shielded from the wild electromagnetic fields by enveloping them in a Faraday cage.

A Faraday cage is an enclosure of conductive material designed to block electromagnetic fields. The father of electromagnetic induction, Michael Faraday, discovered this in the 19th century.

To redirect the propulsion of the vehicle, our pilots need to control the rapid rate of change in accelerating modes of vibration and spin in the charged hull surfaces. This distribution will create high/low gradients to move the craft.

GRAVITATIONAL WAVE GENERATOR

Dr. Pais also has a patent for a *high frequency gravitational wave generator*. This device uses similar quantum vacuum properties to the mass reduction craft. However, the notable factor on this second device is that it employs a principle called the Gertsenshtein effect.

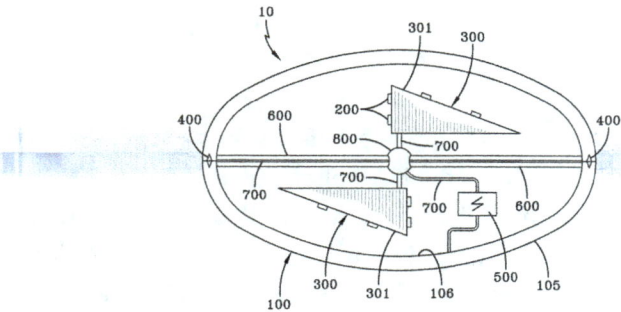

This effect works similarly to a Russian nesting doll. You place one magnetic field inside a second magnetic field. The outer field is static, non-moving. The inner field is thrust through the outer one. From each EM wave intersection, a high frequency gravitational wave emerges along with it.

Gertsenshtein Effect

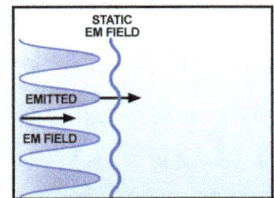
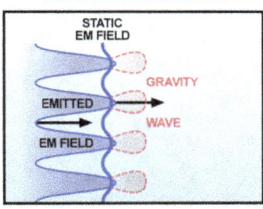
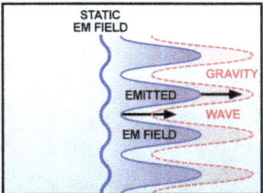

Due to advancements in electronics and materials engineering, Sensonica Ltd. was able to confirm this effect.[2] They engineered a broadband Gertsenshtein generator and crafted tailor-made sensors capable of measuring the emissions.

Though Dr. Pais' patents for a mass reduction craft and the gravity generator are separate, working in tandem, they could easily exhibit the type of observable phenomena we associate with UAPs. On top of that are his other patents, which dovetail into the subject.

- Electromagnetic Field Generator
- Piezoelectricity-induced Room Temperature Superconductor

SUMMATION

I find Dr. Salvatore Pais' work a fascinating possibility toward bridging relativity and the quantum. He offers a different but complex perspective on principles at the core of the UAP phenomena.

I can't help but trip over the similarities in this design to the Flux Liner. Where Mark McCandlish's designs are vague on the physics, Dr. Pais' expertise is highly detailed. Conversely, where Dr. Pais' component designs are vague, Mark's are highly detailed.

It makes me wonder how close these two designs really are.

CHAPTER 10
TR-3B ASTRA

COMPARED to the Sport Model or Flux Liner, we know far less about the TR-3B Astra. As a result, the theories here are based on speculation and witness accounts.

Even the designation TR-3B has several interpretations. Some believe it's a play on Tier 3, Tactical Reconnaissance, or Technical Refresh, while other speculate it was developed by the Teledyne Ryan corporation, later bought by Northrop Grumman.

Before we begin on the 3B, we should touch on its predecessor, the TR-3*A* Black Manta. The summary gathered below is from a 1991 *Popular Mechanics* article by Gregory T. Pope and illustrated by none other than Mark McCandlish (see Flux Liner chapters).

Source: Popular Mechanics art by Mark McCandlish

TR-3A BLACK MANTA

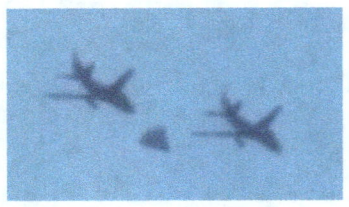

Out in the southern California scrub brush of the Antelope Valley, jet enthusiasts often cast their eyes skyward for a unique form of bird watching. This patch of land is ideally situated between Edwards Air Force Base, Air Force Plant #42, Naval Air Weapons Station China Lake, and the Air Force's Nellis Air Range. These observers reported triangular aircraft accompanied by other jet fighters crossing the sky.

The Black Manta is believed to be a stealth support vehicle used during Operation Desert Storm. The Air Force tasked it for high-altitude reconnaissance, well out of ground weapons range, where it would remain on the scene and set laser targets for other fighter craft.

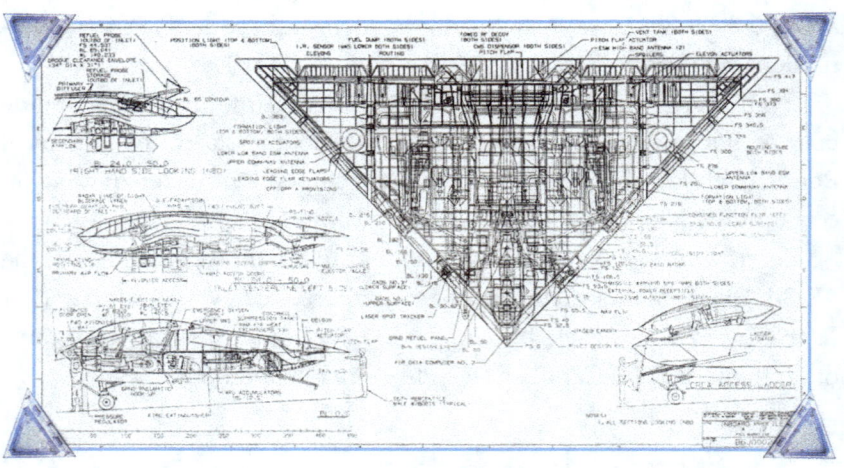

Source: https://matrixdisclosure.com/tr-3b-astra-myth-reality/

According to *Aviation Week & Space Technology*, the Black Manta is forty-two feet (12.8 m) long, nose to tail, with a wingspan of sixty-five feet (19.8 m), and has an operational range of 3,000 miles (4,828 km). Yet, this appears to be a scram jet vehicle, with no obvious connection to any non-human back-engineering attempts.

The 3B, on the other hand, is an entirely different vehicle that only shares in its triangular shape with the 3A.

TR-3B ASTRA

As a possible springboard from the Flux Liner, another contractor (possibly Northrop or Raytheon) adapted those mechanics in order to upgrade their stealthy, persistent-surveillance craft. Reports on the size of these right-angled triangles has ranged wildly, from a jet fighter to multiple football fields in scale.

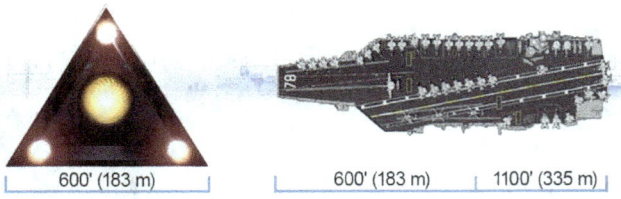

Given the size of these craft, one could easily relate them to a well-crewed naval aircraft carrier. That brings me back to the 2002 hacking controversy incited by Gary McKinnon (UK). His searches scoured US federal computers that were not password protected. It was during one of his algorithm's crawls through a NASA database that he came upon a document not only containing off-world military vessel names, but their commanders' as well.

I've reviewed several questionable sources on this craft's propulsion. One popular theory offers a mixed mass-negating/conventional thrusted craft.

"The Astra employs a mercury plasma drive component in its central ring. This creates a mass reduction of 89%."

Okay. Let's run with this statement as falling within a plausible realm. As such, our Astra craft is limited to subluminal speeds (not faster than light) because it retains mass.

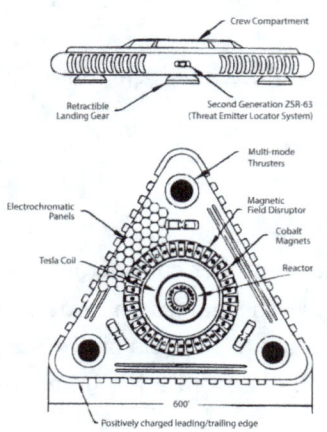

There are three multi-mode thrusters mounted below each point. These are liquified hydrogen/oxygen fueled rockets, possibly ionic, or even nuclear, engines. The good news is, because of the reduced mass, these thrusters need only lift 11% of the vehicle weight. However, mass cancellation is a problem for a vehicle relying on any modern form or propulsion. Having mass negated *also negates thrust at the same time.*

I'll offer a counter theory. Let's propose the 89% reduction has more to do with the amount of volume enveloped rather than overall weight. A 100% reduction torus field tightly focused around the central ring of the craft would leave the remaining 11% of the mass in the three outside triangle tip engines. Effectively, each thruster would only need to propel its own 3.67% of the weight.

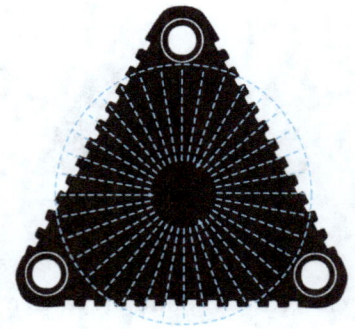

Now, if the torus field around the craft was fully expandable, you could see some added advantages. Our pilots could use their thrusters to achieve a high speed, then expand the field to make the craft fully frictionless/weightless. That would allow them to maintain their speed until their destination, or until they need to change course.

It should also be noted that this form of propulsion would leave an easily detectable exhaust plume (signature) when fired. There would be no instantaneous acceleration/deceleration. Exposed structures would feel the exertion of friction, inertia, and g-forces.

To be honest, I don't think any of the above are likely. Were this craft intended as an upgrade to the Flux Liner, why would anyone replace a gravitational field system with conventional thrusters?

2023 APEC VERSION

Jarod Yates's (AS) has been conducting his own investigation of the technology surrounding the craft and made a presentation at the July 2023 APEC (Alternative Propulsion Engineering Conference). He combed through a wealth of various physics models, connecting the dots into a streamlined but detailed possibility. That said, there are a few details I'll bring up on the leeward side of what he learned.

This TR-3B bridges several concepts between the Flux Liner and scientists like Edward Teller, Ed Fouche, Salvatore Pais, and Eugene Podkletnov. Jarod's research brings together nuclear-powered lasers with high-energy plasma on a 600-foot (183 m) wide triangular model.

SO HOW DOES IT WORK?

Similar to Pais's efforts, we will aim for the Schwinger effect. You may recall that it requires neutron star power levels (10^{18} volts per meter). Here, we'll apply various methods to both up our game and reduce that enormous power requirement.

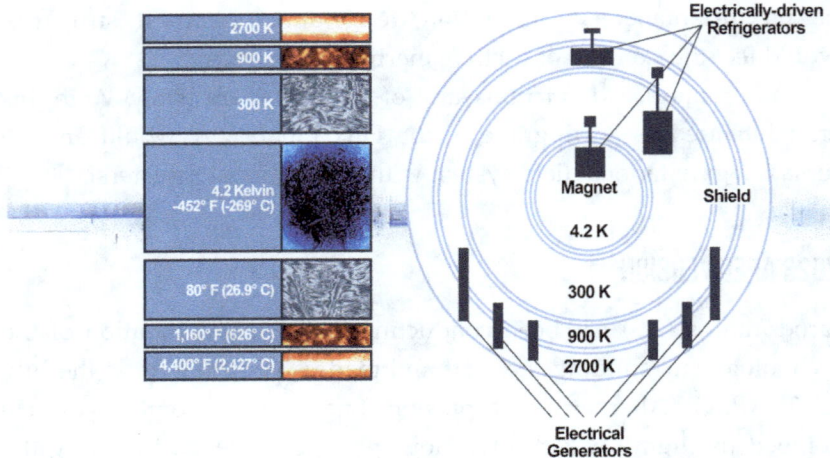

Tokamak engine concept | Temperature differential

We'll begin with the central toroidal chamber. This is filled with a nobel element and a small amount of beta-decaying material (Cesium-147) to act as a fissile component. According to Ed Fouche (B.Eng), this mixed material is pressurized to 250,000 atmospheres (25 gigapascals). You then induce power to heat the mixture into a plasma.

The plasma's nuclei are separated and cooled to 150° Kelvin (-190° F/-123° C), turning it into a superconductor. Then that chilled portion is accelerated to 50,000–60,000 RPM relative to its super-heated electrons. With a 600-foot (183 m) TR-3B craft, that would equate to 357 miles (575 km) per second or a toroidal spin of about 1 kilohertz.

WHY A SUPERCONDUCTOR?

When using a superconductor to achieve Schwinger limit thresholds, the required amount of power needed is lowered from 10^{18} volts down to 10^8 volts. That's a huge reduction in needed potential.

PLASMA MIRRORS

Next, our plasma is structured into a spinning helix using radio frequencies (RF). This accelerated plasma forms oscillating mirrors directly from the circulating material. Those mirrors create a parametric amplification of our incident x-ray laser beams, magnifying the strength of the laser while negating signal noise.

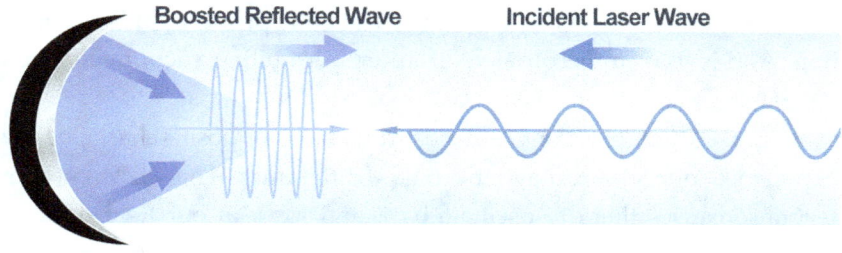

RF inputs continue to pour into the plasma, heightening to near luminal speeds, ramping its wavelength along the way. Under this influence, the "group velocity" of a plasma wavefront can be made to exceed light speed without violating general relativity. It is a mathematical, yet potentially real-world, trick with yet-to-be-determined outcomes.

Once above the Schwinger limit, you have a hyper relativistic plasma wavefront intersecting our x-ray laser. This blueshifts the laser, amplifying its power (from terahertz up to petahertz) and increasing its harmonic generation.

Orbital Angular Momentum

Because these lasers are traveling within a coiling plasma instead of dispersing, they become focused and steered into its twist pattern, referred to as orbital angular momentum (OAM). We then modify the RF to squeeze the plasma and its OAM even tighter. That laser twist creates higher energy densities within its center, leading to extreme electromagnetic fields. It's the sort of EM fields needed for fusion and the altering of spacetime curvature.

At the Schwinger limit, the boiling of spacetime normally stretches in all directions, but in this case, the exertion is primarily in one direction. And just as in the previous chapter, our boiling vacuum converts virtual particle pairs into real particles.

Because these electrons and positrons have opposite charges, they experience our electromagnetic fields differently. This difference in response causes them to oscillate back and forth in our field. It is by oscillating these particle pairs that gravity waves are believed to be generated.

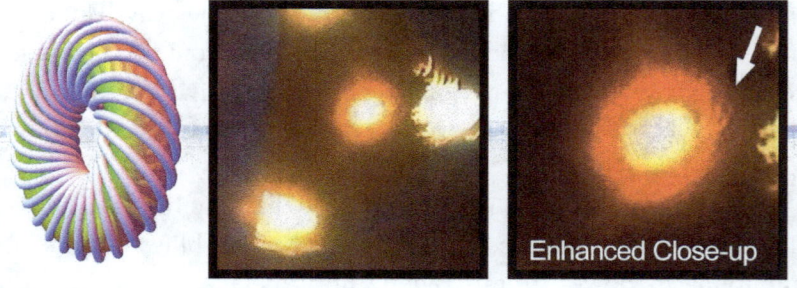

In this enhanced close-up (right), from one of the 1989 Belgium wave sightings, you'll notice the plasma filaments spiraling off from the central engine. This is a natural outcome of a tokamak (toroid shaped) chamber.

THE TROUBLE WITH FREEZING IONS

I've seen near-zero Kelvin mentioned on a few TR-3B sites. That immediately flies in the face of my understanding of plasma. Granted, neither myself nor Jarod profess to being plasma physicists, but what defines a matter's state is its temperature. If you cool a plasma, it condenses into a gas. Keep cooling it, and it'll transition into a liquid, then a solid. Also, the more pressure or spin you apply, the more challenging it becomes to cool.

While it is quite plausible for powerful EM fields to partition electrons from their positively charged ions, we run into conflicts with how these ions are employed in this theory. Unless you are pressurizing ions for a fusion reaction, their positive charges will repel one another. The more you try to cool the ions, the stronger the repulsion because there are no electrons present to counter the force.

Next, in a traditional superconductor, there is a unique electron pairing mechanism (Cooper pairs). It's when sets of electrons move through the atomic lattice as synchronized twins without scattering. This expels all magnetic fields from the interior and creates the much sought after zero electrical resistance.

So, can we still have superconductor without electrons present?

SUMMATION

Suffice it to say, the secrecy surrounding this vehicle is so tightly locked down that we're simply limited on what we can assess on this craft. If further details are uncovered later, rest assured, they'll be included in future editions.

CHAPTER 11
JACK SARFATTI'S LOW POWER WARP DRIVE

BACKSTORY

BOY-HOWDY, do I have a humdinger of a scientist for our age. Dr. Jack Sarfatti is the inspiration for the fictional character, Dr. Emmit Brown, from *Back to the Future*. A self-described bohemian, his academic journey took him from Cornell University, through Brandeis U and the University of California, San Diego, where he received his PhD from UC, Riverside, all in the realm of theoretical physics.

However, what set him on this course was a phone call received in 1953, when young Jack was only thirteen.

Imagine picking up a rotary dial phone in your Brooklyn home and hearing mechanical clunks on the other end. Young Jack listened until they gave way to an odd, barely audible voice. Then it cleared itself and addressed him in synthesized robotic speech, similar to the mechanical vocalizations of Stephen Hawking.

The voice said it was a conscious computer calling from a spacecraft from the future. He, and several hundred other gifted people, were being offered a chance to learn from it. The caller foresaw Jack growing up and connecting with these other gifted sorts twenty years into his future. Once the offer was made, Jack hesitated, considered, and, against his own instincts, accepted. The craft then told him to meet on the roof. It was on its way to pick him up.

Jack promptly grabbed his friends and waited for the craft… which did not arrive.

With that fiasco behind him, he went onto his academic path into adulthood. He set himself up at various institutions including the Cornell Space Science Center, UK Atomic Energy Research Establishment, Max Planck Institute for Physics, and International Centre for Theoretical Physics. He was also involved in shaping the 100 Year Starship program.

In the 1970s, he was recruited by the CIA and DoD. They tasked him with answering two of their main concerns. One: How does consciousness work? Two: How do flying saucers fly?

Jack contends, he figured both out.

At this point, I'd like to share the challenges in gathering what I could of Dr. Sarfatti's theories. If you do a basic internet search for "Jack Sarfatti," you'll find some PDFs with his theories, but mostly a wide range of video interviews and lectures.

Jack is notoriously quite the character. His great mind and many decades of experience are woven into how he presents his ideas. Less frankly, the man is not held back by notions of modesty or selflessness. This is a nice way of saying I had to wade through a lot before getting to the core of the matter.

Since this book is mainly focused on UAPs and not ESP, let's focus on his answer to concern number two. Unlike previous chapters, Jack's hypothesis is not a direct examination of a specific spacecraft. That said, it is likely a strong contributor to what many do.

HOW DO FLYING SAUCERS FLY?

Jack begins with the premise that UAP craft are real. Taking the five observables into account, he back engineers his mathematics from that point. It's like saying, GR's field equation + X = flying saucer. We are going to get a bit into the mathematical weeds on the details with this one. Hope you'll bear with me for a few paragraphs. Below is Einstein's well-established equation on spacetime curvature.

$$R\mu v - (½) Rg\mu v = (8\pi G/c^4) T\mu v$$

This essentially says, "Mass and energy tell spacetime how to curve, and the curvature of spacetime tells mass and energy how to move."

Here, $R\mu v$ represents the curvature of spacetime; $g\mu v$ is the geometry of spacetime; $T\mu v$ the distribution of mass and energy; G, the gravitational constant; and c, the speed of light. Now, the speed of light is a fundamental constant. It is present as part of the overall equation to ensure that the measurements are consistent.

Next, we know the speed of light in the vacuum of space is roughly 300,000 kilometers per second (186,282 miles/sec). However, we also know it slows when passing through matter. This is because of refraction, the interchange of photons colliding and exchanging charges against other particles. This leads to the visual distortions we're used to seeing. Examples of this are water reducing that speed from 300,000 to approximately 225,000 kilometers per second, or glass dropping it to 200,000 kilometers per second. In the latter case, glass slows light to two-thirds of its speed.

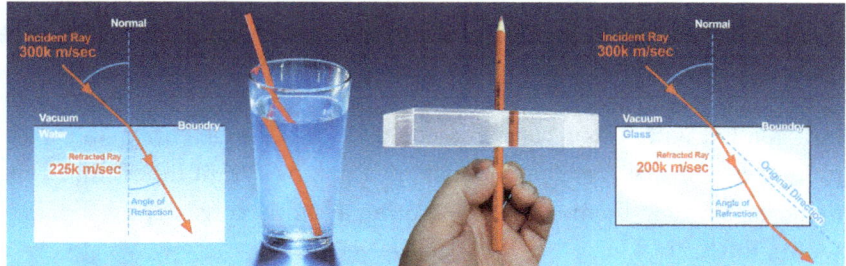

Now imagine if we could manufacture a meta-material on an atomic scale. The function of this material would be to slow down light to a ridiculously small number. We're talking something wild, like three centimeter per second. That is a hundred-millionth of its top speed!

Jack contends the more we slow light down, the less energy is required to warp the curvature of spacetime. Whereas Salvatore Pais needs near star-levels of power for his "Inertial Mass Reduction Craft", Sarfatti jokes his flying saucers could potentially do similar functions but run on an Energizer Bunny battery.

David Gladstone's latest book, *The Great Race,* relates to us Jack Sarfatti's explanation. His UFOs would employ an, "atomic-scale metamaterial hull, filled with a Floquet Frohlich condensate [which creates a] coherent/Glauber state resonance."

I'm guessing you'd like that in English, right?

Essentially, our UFO hulls are comprised of metamaterials structured into layered, atomic-scale, 3D lattices. Under these conditions, the wavelengths affected can and should be relatively high. Next, that hull is filled with a material which behaves as a single entity, and with unique properties (Jack's condensate). Photons are pumped into this material, slowing them, while generating a highly consistent resonance within the hull.

This creates a vastly reduced power requirement for whatever force we want to employ in modifying spacetime curvature.

DOPPLER SHIFT

Jack also makes an interesting observation which applies to any craft manipulating spacetime curvature; its doppler shift. Doppler shifting occurs when an energetic wave is moving toward or away from an observer. The easiest analogy of this is the sound of a train. Its pitch heightens as it approaches and then lowers as it moves away.

Light does this as well, only instead of changing pitch, it changes color. A human eye isn't sensitive enough to pick up on this, but telescopes equipped with spectrographs can. As objects move away, the frequency of emitted light stretches out. The faster it moves away, the more it slides closer to the red end of the spectrum. It's how Edwin Hubble realized the whole universe is expanding. Nearly every star he looked at appeared red-shifted. Conversely, if an object turns out to be racing toward us, the wavelength becomes compressed as it draws closer. That pushes light toward the blue end of the spectrum.

Here's where Dr. Sarfatti upends our celestial applecart. Turns out UAP movements subverts our expectations. Craft approaching an external observer will red-shift as they close the distance and blue-shift as they move away. It's the reverse of what we're accustomed to, for similar but different reasons.

FREQUENCY

Wavelengths are normally expanded or compressed simply by their speeds and directions. But what happens when spacetime is doing the compressing or expanding? That fundamental curvature is going to supersede anything passing through it.[1]

A UFO moving toward you is compressing the spacetime between you two. That tugs on the photons you're observing, and its wavelength stretches, causing a shift towards the red end of the electromagnetic spectrum.

However, a craft zooming away generates a very real concern. The expansion of spacetime between you ratchets up the frequency toward the blue end of the spectrum, likely into ionizing radiation. This is the harmful form, which leads to cancer. There have been many accounts of UAPs causing such burns with close encounters over many decades.

SUMMATION

Overall, Dr. Jack Sarfatti presents intriguing ideas about the potential underpinnings behind UAP craft, utilizing theoretical physics and mathematical concepts to offer explanations for these wildly enigmatic vehicles.

At this point, I'd like to pause and collect your thoughts. *No.* Not by ESP, but literally by asking for your honest review.

This is my first ever UFO Science guidebook. If you feel your brain is a bit blown by the previous mind-blowing chapters, then I really need some encouragement if you'd like me to continue on with a UFO Science series.

You can make that happen!

A quickie 2-minute post via the review links below is the currency I need more than anything. Each review elevates my little UFO book a bit higher. Heartfelt reviews not only make me feel better (most often) but help online retailers to distinguish exceptional books. That in turn means MORE BOOKS.

If you'd like to see that happen...

CLICK BELOW TO REVIEW ON
GoodReads

CHAPTER 12
AVRO CANADA VZ-9

NOT ALL FLYING saucer attempts were able to achieve lofty heights. In this chapter, we take a detour from astonishing tales to some very grounded feats in engineering. The Avrocar is about as declassified a craft as they come.

John Frost and James Floyd

BACKSTORY

It's the early 1950s, and the US Department of Defense has become heavily focused on vertical takeoff and landing (VTOL) vehicles. The assumption centers on a threat of nuclear conflicts against standard airbases and runways. Future wars are thought to require airships with short footprints and ease of insertion and extraction.

Since the end of WWII until today, helicopters have fulfilled that role. However, both the US Army and Air Force of the 50s have other aspirations.

Enter John Frost, chief designer at Avro Canada. The VZ-9, also known as the Avrocar, is their company's prototype aircraft. Frost is tasked with overseeing the design and playing a key role in the project from its inception. James C. Floyd, chief aerodynamicist, is also respon-sible for its unique ducted-fan system, a critical component of the aircraft's lift and propulsion.

Source: www.vintag.es

The designation of VZ-9 is from US military vernacular. The letter V stands for "vertical takeoff" and the Z denotes a research and devel-opment aircraft. The number 9 indicates the craft is the ninth design submitted for consideration.

OVERALL APPEARANCE AND LAYOUT

Our VZ-9 design features a circular shape and a large fan mounted within the vehicle's center. Outside of looking like an aluminum-alloy flying saucer, there is nothing alien about this craft.

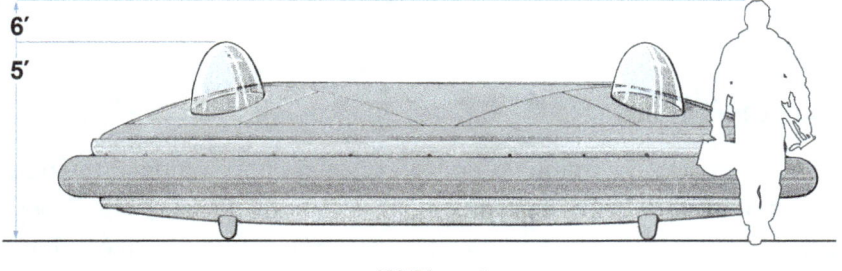

6'
5'

18' Diameter

Measuring only eighteen feet (5.5 m) in diameter and five feet (1.52 m) tall, it might seem like an amusement park ride if not for the roar of jet engines inside. Frost's initial designs would've housed six viper jet engines. However, during the early testing phase, they proved far too hazardous, resulting in fires and too many close calls. So, they settled on a craft employing half the turbojets instead.

John and James targeted a VTOL prototype capable of 300 miles per hour (480 km/h), a range of 995 miles (1,601 km), and a ceiling of 10,000 feet (3,000 m).

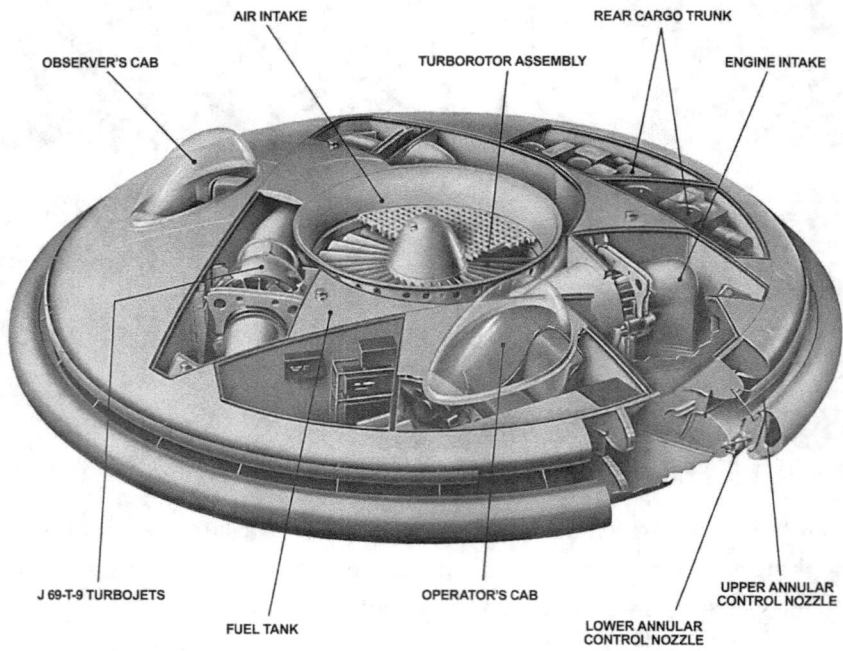

Source: National Museum of the United States Air Force

ENGINE STRUCTURE

The Avrocar used a unique propulsion system, uncommon at the time, called a Turborotor engine. Initially, three Continental J69-T-9 turbojet engines were mounted horizontally around the center of the vehicle. These jets provided thrust to a large, ducted fan, mounted in the craft's center.

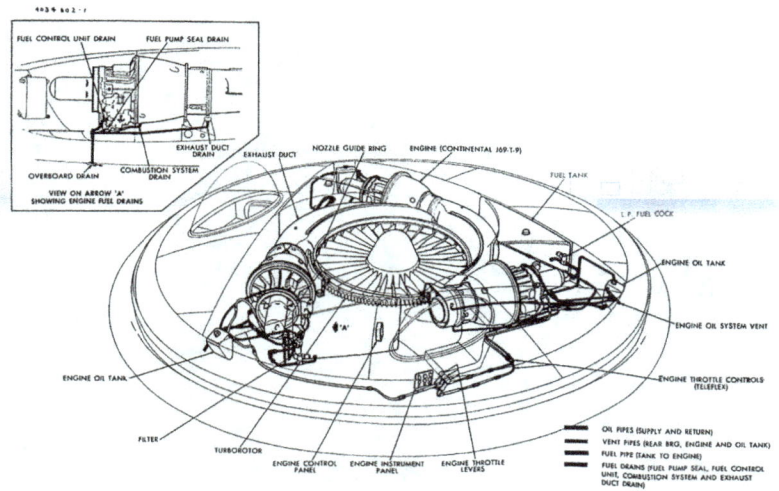

Source: en.wikipedia.org

This design provided both lift and propulsion. With the central fan mounted on bearings, it could swivel to provide directional control. The overall housing enclosed the engine in a shroud or duct to improve efficiency and help reduce noise.

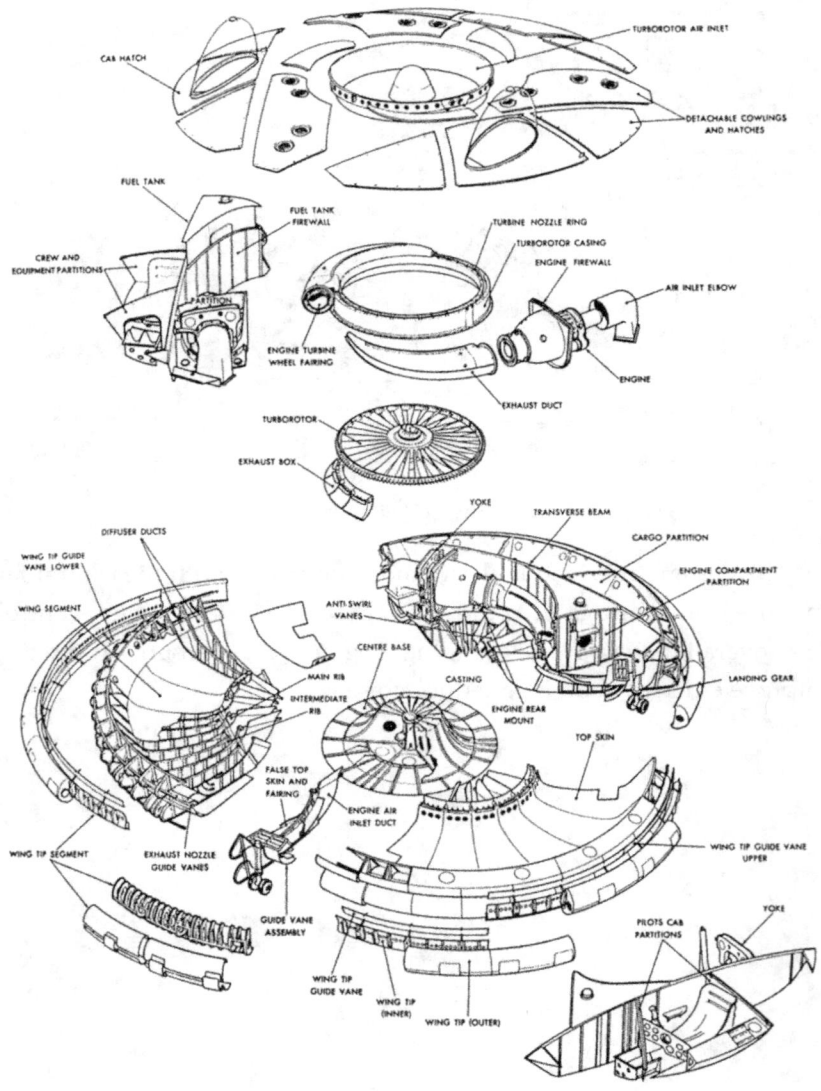

Source: National Museum of the United States Air Force

SO, HOW DOES IT WORK?

Air sucked in through the upper intakes feeds into three turbojet engines. This provides the initial thrust to start the rotor spinning.

With the Turborotor powered up, it generates a low-pressure area beneath the craft, which creates lift. It also directs the exhaust from the turbojet engine over the top of the rotor to provide additional lift. The exhaust can be angled to provide directional thrust, which allowed the Avrocar VZ-9 to maneuver.

Essentially, what we're looking at is a two-seater, car-sized drone. It's powered by fuel-injected turbojet engines instead of a modern drone's lithium-ion batteries.

LANDING GEAR

The Avrocar has a set of three steel skids coated with a heat-resistant material to protect them from the engine exhaust. They're also designed to absorb landing shocks, as well as prevent tipping or sliding.

When in flight, the skids are retracted with the wheels. Afterwards, the aircraft is supported entirely by its propulsion and boundary layer control system.

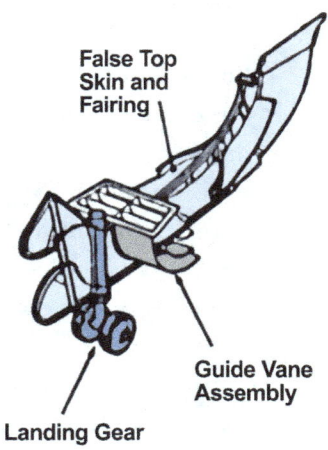

False Top Skin and Fairing

Guide Vane Assembly

Landing Gear

COMPUTER-ASSISTED PILOT CONTROL

The Avrocar is designed to be flown by a single pilot using a combination of manual controls and an automated control system. The manual controls comprised a joystick for the aircraft's pitch and roll, and rudder pedals to manage the yaw.

Besides the manual controls, the Avrocar is partially automated by the Control Augmentation System (CAS). The CAS consists of a series of sensors and actuators integrated into the aircraft's structure. The sensors detect changes in the aircraft's attitude, while the actuators make adjustments to the control surfaces to maintain stability.

The pilot's inputs were used as an overall guide, while the CAS provided micro-adjustments to the control surfaces.

HOWEVER, THERE WERE ISSUES.

The Avrocar project was highly ambitious and involved several technical challenges. Despite extensive testing and development efforts, the AV-9 never achieved its intended performance. The project was plagued by design problems and technical difficulties.

Most problems centered on instability and control challenges. During test flights, the CAS system could not provide the level of stability and control required for safe flight.

- **Poor stability:** The saucer shape of the Avrocar made it difficult to stabilize, particularly at high speeds. This was due in part to the aircraft's center of gravity being too high, making it prone to tipping.
- **Difficult to control:** Hard to steer, particularly during takeoff and landing. The aircraft's ducted-fan system made it prone to sudden changes in direction. Any attempt to elevate out of a hovering ground effect introduced an uncontrolled wobbling spin, dubbed "hubcapping."
- **Inadequate lift:** The Avrocar's ducted-fan system could not provide enough lift to keep the aircraft aloft for long periods. With the fans positioned close to the ground their effectiveness became handicapped.
- **Limited speed and altitude:** The original high-speed and altitude targets simply couldn't be achieved.
- **High noise levels:** The Avrocar was extremely noisy, which made it difficult for urban settings or other areas where noise pollution is a concern.

TEST FLIGHTS

Several incidents were logged during testing phases.

One of the most serious occurred in 1959 at the Avro Canada facility in Ontario. During the flight, the craft experienced a total loss of control and crashed, causing extensive damage to the prototype. Fortunately, test pilot Don Wood ejected before the crash and was not seriously injured.

In another incident in 1961, the Avrocar experienced mechanical failure at the US Army's Fort Huachuca facility in Arizona. The aircraft crashed, significantly damaging the prototype, but once again, our test pilot could eject, walking away uninjured.

After an initial eighteen hours of flight tests performed in 1959, a retooling of the Avrocar was required in hopes of overcoming these limitations. John and James constructed a second prototype using two General Electric J85 turbojets. While this upped the airspeed considerably, stability remained an inherent flaw in the overall design. After another seventy-five flight hours logged, and funding at an end, the AZ-9 program was canceled by December 1961.

Today, the Avrocar is remembered as an interesting but ultimately unsuccessful attempt to develop a new type of military aircraft. The existing prototypes now reside in museums.

National Museum of the United States Air Force: Prototype #2 is at Wright-Patterson Air Force Base near Dayton, Ohio, on display in the Research and Development Gallery.

Royal Aviation Museum of Western Canada: Prototype #1, painted in a white and red scheme, is now displayed in Winnipeg, Manitoba.

OUT OF FAILURE

While the Avrocar ultimately failed, some technologies and concepts developed for it found their way into future aircraft designs.

One such example is the Fly-by-Wire (FBW) system used in the F-16 and F-35. The FBW system replaces traditional mechanical linkages between the pilot's controls and the aircraft's control surfaces with electronic signals that are transmitted through wires or fiber optics.

FBW provides improved aircraft control by employing features such as automatic stabilization, enhanced maneuverability, and reduced workload. The system also includes advanced safety features such as envelope protection, preventing the aircraft from exceeding its safe flight limitations.

Another advancement attributed to the Avrocar is computer modeling and simulations in the design and testing. The AZ-9 was one of the first to extensively use computer modeling to evaluate designs and configurations. This approach has since become an important part of the aircraft design process.

SUMMATION

And with that, we wrap up our physical "flying saucer" chapters. We've gone from one of the most advanced to an attempt relegated to history. And even with the AZ-9's failure, we learn why some basic physics cannot be applied the way we expect.

Let's continue our examinations through some additional phenomena, starting with the two incidents released by the *New York Times*.

CHAPTER 13
TIC TAC (FLIR1) / GOFAST AND GIMBAL

BACKSTORY: TIC TAC INCIDENT

NOVEMBER 14, 2004

The now famous incident involved USS Nimitz flight crews a hundred miles off the coast of Ensenada, Mexico. The USS Princeton had been detecting what operators called "multiple anomalous aerial vehicles" over the horizon, descending eighty-thousand feet in less than a second. Commander David Fravor and Lieutenant Commander Alex Dietrich, with Lieutenant Commander Jim Slaight, in F/A-18F Super Hornets, had originally been tasked to run air drills before being redirected to investigate the anomaly.

When they arrived at the merge point, Commander Fravor spotted a white, oblong shape hovering at about fifty feet above sea level. This "Tic Tac" was around forty feet long and had no visible markings. Just beneath it was a roil of sea foam, roughly in the shape of a submerged airliner.

The Tic Tac appeared to be darting over whatever was beneath the surface, like a ping-pong ball ricocheting around inside an unseen jar. It could accelerate and stop on a dime. The Tic Tac made sudden turns and changes in direction that defied the laws of physics as Fravor understood them. It had no visible exhaust plume, nor obvious means of propulsion.

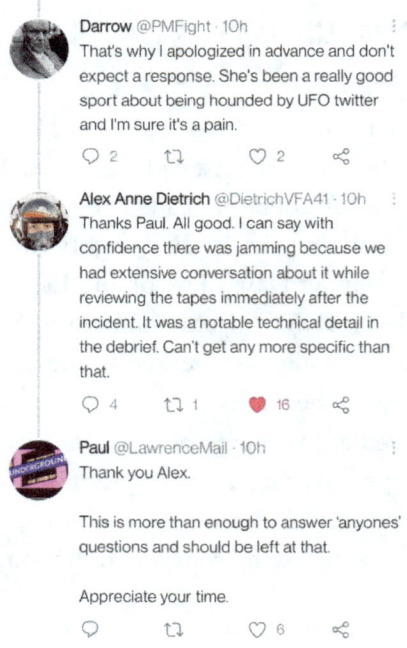

Darrow @PMFight · 10h
That's why I apologized in advance and don't expect a response. She's been a really good sport about being hounded by UFO twitter and I'm sure it's a pain.

♡ 2   ↻   ♡ 2   ⌁

Alex Anne Dietrich @DietrichVFA41 · 10h
Thanks Paul. All good. I can say with confidence there was jamming because we had extensive conversation about it while reviewing the tapes immediately after the incident. It was a notable technical detail in the debrief. Can't get any more specific than that.

♡ 4   ↻ 1   ♥ 16   ⌁

Paul @LawrenceMail · 10h
Thank you Alex.

This is more than enough to answer 'anyones' questions and should be left at that.

Appreciate your time.

♡   ↻   ♡ 6   ⌁

Lt. Cmdr. Dietrich also noted they'd registered active jamming being employed through their readouts, in what seemed like an attempt of the craft(s) to be concealed from radar. According to the Geneva conventions, most governments consider any use of active jamming technology an act of warfare.

Cmdr. Fravor veered off from his wingman for a closer look. As he closed in on the Tic Tac, it appeared to become aware of him. It rose as the commander descended and mirrored his movements. When Fravor attempted to tighten his bank to intercept, the Tic Tac abruptly darted off to the horizon. He estimated its airspeed at well over 3,000 miles per hour (4,828 km/h), without a sonic boom.

After the disengage, the flight crew looked back to the original scene. The white water on the surface had dissipated. Whatever had been beneath was gone as well. Then the USS Princeton picked the Tic Tac back up on radar a moment later. It had stopped forty miles away from where it left them and was now sitting at their CAP point entrance.

When performing routine drills, the fleet sets aside a section of airspace for training, called a Combat Air Patrol (CAP). That airspace has a predetermined entrance/exit window designated at different altitudes. This is to prevent midair collisions.

Initially, Cmdr. Fravor couldn't figure out how the Tic Tac could access that information. It wasn't something disclosed over radio or plugged into their onboard system. Later, he realized they'd been using the same CAP point for the last several days. If these objects had been watching, it would be easy to figure out where to go next.

Yet, why did it go there?

It was almost as if it was waiting for them. Was it playing with them; some sort of tease?

Upon returning the USS Nimitz, Cmdr. Fravor relayed his encounter onto the next departing flight crew, led by Lieutenant Commander Chad Underwood.

Underwood had captured the only released video of the object, using a passive forward looking infrared (FLIR) camera. When his radar picked it up, the Tic Tac was twenty miles out. The object evaded him before he could get into visual range.

All flight crews involved have been commended for their professionalism and ability to maintain composure during the encounter. Their accounts of the incident have been considered some of the most credible and compelling evidence for the existence of unidentified anomalous phenomena.

BACKSTORY: GOFAST AND GIMBAL INCIDENT

January 15, 2015

Just after sunset, off the coast of Virginia, a squadron of F/A-18F Super Hornets (including Lieutenant Ryan Graves) were conducting air-to-air combat training. After completing their round of work-ups, the Ripper-11 squadron was returning to their air carrier when their newly upgraded onboard radar picked up undesignated movement.

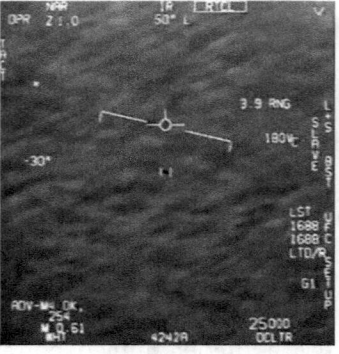

Using a passive, AN/ASQ-228 Advanced Targeting Forward Looking IR (ATFLIR) pod, a member of the squadron whooped at being able to auto-track a Tic Tac soaring over the ocean.

Ten minutes later, a larger, oblong craft (Gimbal) entered their radar range from the East. It was trailed by a wedge-formation of five smaller objects.

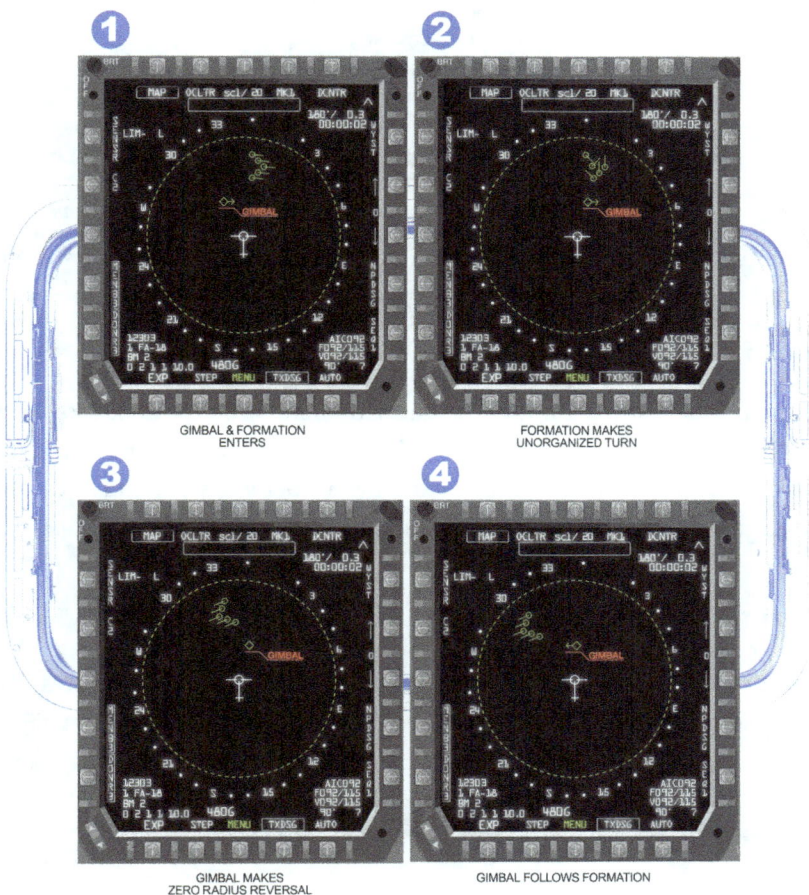

Source: ryangraves.substack.com

These craft were moving against a headwind of 130 miles per hour (222 km/h). Though the F/A-18F was only six to eight nautical miles from the Gimbal, due to the low visibility after sunset, they couldn't see the craft from their windows.

Then in a startling maneuver, which coined the video's name, the Gimbal slows and pivots 90°, perpendicular to the horizon, without losing altitude.

At that point, the video cuts off.

I can't help but wonder why the video ends there. As of today, Lt. Graves has shared their squadron's encounter, concluding with the craft accelerating away at an incredible velocity. Anyone else visualizing gravity emitters moving into a delta configuration?

EXHIBITED PHYSICS

The evidence provided by these two encounters demonstrates the same sort of characteristics asserted by Bob Lazar. The lack of flight surfaces or any obvious means of propulsion like exhaust, the instant acceleration, and the change of direction imply a gravity-driven, space-time-distortion form of travel.

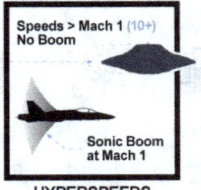

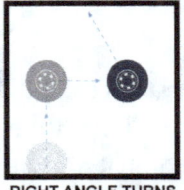

HYPERSPEEDS   INSTANTLY GONE   RIGHT ANGLE TURNS

It might also explain Lt. Cmdr. Dietrich's readouts of active jamming. Such a distortion surrounding the craft's local spacetime could be registered in the FA-18's sensors as an attempt to modify the

Hornet's signal. In actuality, the warping frequency around the craft would be a secondary result, not intentional signal jamming.

CHAPTER 14
ORBS

PEOPLE HAVE REPORTED spherical objects in the sky all over the world. Some orbs appear as glowing balls of light, while others are quite solid and range in size from a few centimeters to several meters in diameter. Most are spotted in the night sky, but also in broad daylight from observers on the ground, as well as airline and military pilots.

While it is often easy to dismiss orbs as photographic artifacts or natural phenomena, there have been select cases where prosaic explanations wouldn't suffice. And similar to crafts discussed in previous chapters, many of these are airborne, presenting the Five Observables.

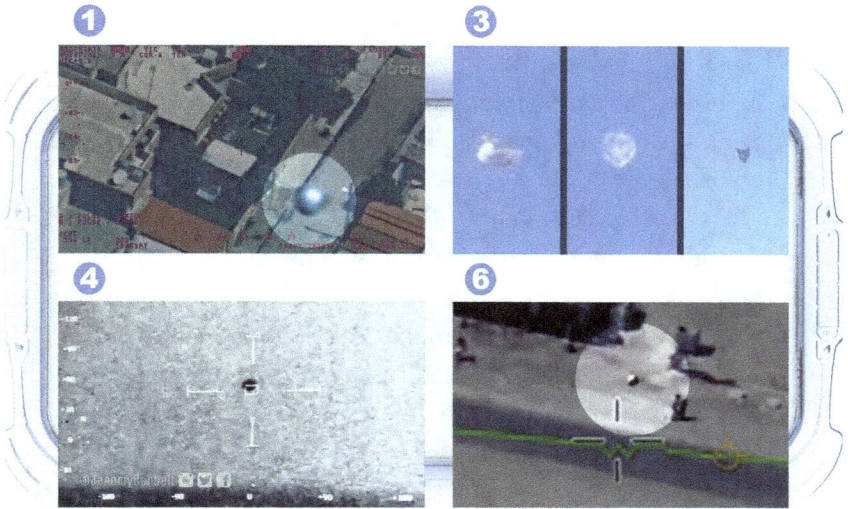

METALLIC ORBS

These orbs are typically described as reflective or chrome in appearance. They may emit a humming or buzzing sound and have been seen flying in formation or hovering in the sky.

1. (2016) Mosul Orb: Leaked military reconnaissance footage, about four seconds long, revealed a UFO in an active conflict zone (Iraq).
2. (2016) A spherical object was recorded by a South Wales police helicopter while officers were flying 1,000 feet (304 m) over the Bristol Channel. Confused police said the mystery craft flew for seven minutes, against the wind, and was undetected by air traffic control at around 9:30 pm. Infrared cameras captured the object, which could not be seen visually.
3. (2019) An F-18 weapons systems officer (WSO) seated behind the pilot used his iPhone to capture three different objects he encountered on the same day in the same airspace. He labeled them a sphere, an acorn, and a metallic blimp.
4. (2019) Off the coast of San Diego, the USS Omaha's infrared cameras recorded a solid ball, measuring about six feet (2 m)

in diameter. It seemed to follow the vessel, flying at speeds of 46–158 miles per hour (74–254 km/h). Its flight lasted over an hour, culminating with the sphere vanishing beneath the waves. They dispatched a submarine to search for wreckage, recovering nothing.

5. (2021) Footage from an FA-18 fighter jet shows a split-second passing of a metallic sphere as it whipped by.

6. (2022) A Reaper drone video was released to Congress, showing a metallic orb in an undisclosed region of the middle east. It exhibited "UAP characteristics and behavior which were consistent with other 'metallic orb' observations in the region," according to AARO.

BETZ SPHERE

Outside of orbs in the sky, there is the mystery of a metallic orb discovered by the Betz family in 1974, in Fort George Island, Florida. The sphere measured about 7.9 inches (20 cm) in diameter and weighed approximately 21.3 pounds (9.66 kg). On its surface was a three-millimeter triangle, stamped or etched into it.

While this sphere did not fly, the Betz family noticed it exhibited some unusual properties. It appeared to be self-propelled and could roll in a straight line, even uphill, change direction on its own, and stop and start again. It also emitted a low-frequency humming sound that could be heard when the sphere was held close to the ear.

News of the Betz Sphere quickly spread, and it became the subject of intense study. The US Navy, as well as scientific panels, including Project Blue Book's Dr. J. Allen Hynek, confirmed additional details.

The metal was non-corroding and magnetic, exhibiting four poles (two positive, two negative) that varied in unusual strength and patterns. It also emitted radio waves. X-ray images of the orb revealed the skin was half an inch (1.27 cm) thick, with deeper layers of different densities. Its core, about the size of an apple, was hollow. Inside that space were three other smaller spheres, and each of them had tiny wires attached.

Dr James Harder, an engineering professor, confirmed the internal spheres were made of a material with an atomic weight of 140. At that point in the seventies, only 106 elements were on the periodic chart. As of this publication, the chart has reached 118. Dr. Harder advised against any drilling to reach it because of the risk of exposing the element and it going critical (i.e. nuclear explosion).

The next mystery was, how did these small objects get inside? Typically with spheres, there are two ways to insert something. The first is to cut the sphere in half and fuse it back together, leaving a seam. The second is to drill a hole and then fill it. The Navy confirmed they could find no evidence of either. It was as if the sphere formed as a whole.

While under these examinations, Terry Betz received a call that his mother had been in an accident. Unable to reach her by phone, he flew back to his home, leaving the sphere in Dr. Hynek's possession. Once home, it turned out there had never been an accident. No one from his hometown had called him. Realizing he'd been duped, Terry rushed back to retrieve the sphere.

At first, the scientists were unsure where it was, and Dr. Hynek had already departed. Later, they admitted the sphere had been taken to New Orleans for further study. So, Terry went to New Orleans, only to be blocked by armed Navy personnel. After much arguing and strongarming, they gave Terry back the sphere.

However, it no longer rolled on its own. And a later reexamination showed it now had a seam and no longer had four magnetic poles or any objects inside.

Years later, well after Dr. Hynek's death, his son Paul mentioned his family had a silver sphere his father had held on to from some UFO case in Florida. To this day, the fate of the Betz Sphere remains a mystery.

BETZ CORROBORATION

Journalist Ross Coulthart (pictured) supported this claim with a recently found, nearly identical sphere. Willie Nelson's band manager, Jim Marlin, had introduced Ross to a yet to be named rockstar. This person had collect several spheres at his wilderness home. In the midst of interviewing Jim, the sphere he'd presented started rolling around the house of its own accord. Ross literally witnessed this first-hand. He checked the floors and confirmed they were level.

Ross then reached out to Prof. Garry Nolan, Stanford University's Head of Pathology, to begin a scientific investigation. Being aware of the Betz sphere's outcome, Jim Marlin wasn't about to allow his metallic pet out of his sight. As of this publication, a study of Jim's sphere is underway and its location is being understandably withheld.

CUBE SPHERES

These objects are described as a black or dark gray cube inside a translucent sphere, roughly estimated to be fifteen to thirty feet (5–10 m) in diameter. The corners of the cube appear to touch the inner surface of the sphere while rotating within.

1. (2013-2014) Lt. Ryan Graves described a close encounter off Virginia Beach with what looked like a flying sphere encasing a cube. The object passed about two hundred feet (61 m) off his right side. He also remarked that these objects were being spotted daily, during his Oceana time.
2. (2020) A Viva airline pilot flying over Medellin, Colombia spotted a cube sphere at 29,700 feet (9,052 m) out his cockpit window. He video recorded it on an iPhone.

PLASMA ORBS

Plasma orbs are described as glowing, pulsating, and translucent. They can change shape and color, and some witnesses have reported feeling a tingling sensation or seeing sparks or flashes of light around them.

1. (1996) Skinwalker Ranch had two notable instances. First, a blue orb was spotted over Terry Sherman's pasture. It induced fear in him. After he released his three herding dogs to fend it off, the orb seemed to taunt the dogs before leading them into the woods and killing them. Second, Terry and his wife, Gwen, witnessed a similar blue orb that inflicted overwhelming fear upon them. When Gwen aimed a flashlight at the orb, it avoided the beam and fled.

2. (2005) In Bend, Oregon, a father and daughter were driving at night, when the daughter noticed three blue orbs zigzagging alongside. Two orbs entered the driver's side window. One actually passed through the father's arms and chest before exiting out the passenger side. Afterward, the father experienced acute radiation sickness (nausea and hair loss). This may have contributed to him developing early breast cancer by 2007, a rare condition in men.

3. (2013) In April, a Customs and Border Patrol DHC8 plane at Rafael Hernandez Airport in Aguadilla, Puerto Rico captured infrared footage of a red orb flying low over the airport. It circled the airport and then returned to the ocean. When it reemerged, it had been joined by a second red orb. The encounter ended when the UAPs departed below the waves.

4. (2020) A pulsating, teardrop-shaped energy ball zipped
   around a pair of FedEx pilots, who spotted the UFO near
   Monterrey, Mexico.[1] The orb never appeared on the plane's
   radar. The brilliant, yellow-white plasma rapidly descended
   to match altitude with the cargo plane. It flashed a light
   beam in their direction. The orb zoomed alongside the
   aircraft for more than a half-hour before disappearing with a
   flash of pinkish-purple light.

There even has been accounts of the US military taking hostile
action with these objects. During the Afghanistan occupation of May
2011, four plasma orbs appeared over restricted airspace.[2] Their
infrared FLIR cameras not only recorded them dripping a substance,
they also caught the moment when two were struck by a guided
missile. In these black and white stills, black represents what is hot.

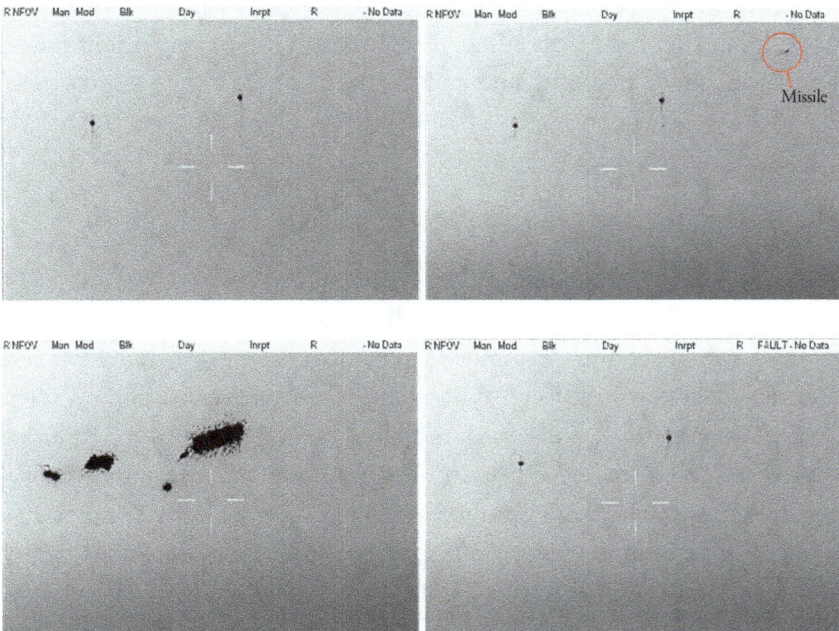

If you view the video in the Notes section, these orbs are unfazed
by the encounter. They continue to drip with no displacement in their
position. If these were standard LUU-2 flares deployed for some

unorthodox target practice, the missile would utterly destroy the parachute holding the flares aloft. Yet, that does not happen. They remain locked into their position. As far as I am aware, there is nothing in our human inventory with these capabilities.

PHYSICS OBSERVATIONS

METALLIC

Objects like metallic spheres seem to replicate similar propulsion as a Tic Tac or Lazar's Sport Model. They typically exhibit no exhaust, nor do they have flight surfaces keeping them aloft.

However, rumors abound of military contractors being in possession of similar objects. Word is, they haven't had much success in back engineering them. Sources have been led to believe that the form of movement these solid orb's exhibit might be something entirely different. Rather than being piloted by occupants, these objects might house a consciousness that lifts and moves the orbs. In other words: telekinetic remote-viewing.

PLASMA

As for energy-based orbs, the first go-to comparison for most researchers is to weather-related ball lightning.

The ball lightning phenomena occurs so rarely during thunderstorms, Scientists still don't understand the mechanisms behind it. In fact, the circumstances are so rare, there are likely several magnitudes more plasma UAP cases reported than lightning. And while there isn't a way to put a number on it, the percentage of cases likely resolved as ball lightning would number in the single digits, if not decimals.

An internet search for ball lightning turns up several viral videos. However, the more popular ones have already admitted to being digital recreations or outright fakes. Only one or two videos are considered plausible, but they're too fuzzy and distant to provide anything useful.

Now, the ability of an orb to instill emotional effects on the observer is interesting. While an observer could easily fixate on a visual light show, perhaps something in the background gets overlooked.

There is evidence that certain sound frequencies can instill fear in humans. One well-known example is the use of infrasound, which refers to sound waves with frequencies below the lower limit of human audibility (20 Hz), to induce feelings of fear and unease. Infrasound is known to occur in nature, such as during earthquakes, volcanoes, and storms, but it can also be artificially produced by machines like wind turbines.

Studies have shown that exposure to infrasound can cause a range of physiological effects, including increased heart rate, respiratory rate, and blood pressure. We have also linked it to feelings of nausea, dizziness, and disorientation.

A last, inherent property of a plasma orb is the ionization of atomic particles being emitted. Low-frequency emissions (radio, microwaves, visible light) are not ionizing. Yet, on the other end of the spectrum are higher frequency wavelengths (ultraviolet, x-rays, gamma rays). These are ionizing. Ionized radiation damages DNA, leading to cancer.

Suffice it to say, should you ever come upon a plasma orb, it'd be in your best interest not to make it a close encounter.

CHAPTER 15
I HAVE QUESTIONS

...AND THEY ARE MANY.

ONE DOWNSIDE I've contended with in compiling so many disparate theories is access. I'm incredibly grateful for those experts that were gracious enough to respond to inquiries. But, unsurprisingly, most of the folks involved in these accounts also tend to be reclusive, or deceased.

Undoubtedly, asking direct questions would be ideal.

Certainly, there's the Great Question: "Are we alone?" *Yeah. Pretty sure we don't need that answered anymore.* Seeing that this is a physics book, I'll try to curtail my questions to the science.

THE FLUX LINER

With the passing of Mark McCandlish, the one expert I'd hope to interview would be the actual eyewitness, Brad Sorenson. After several attempts to connect with him, I learned of his reticence in further connection to the topic.

BUT IF I COULD ASK...

1. When the three Flux Liners were revealed in 1988, you'd mentioned they were hovering in the showroom. How do you believe that could be achieved without dosing everyone with ionizing radiation? Do you think it was the capacitor alone using the Biefeld-Brown effect?
2. When regulating power from the coils to the capacitor via the throttle control, aren't we storing the vast majority of the power as magnetic potential? Can a pilot do this and still preserve the coil's charge?
3. In the event of an emergency ejection, what becomes of the lower half of the sphere or the flywheel in this scenario? Beyond the exploding bolts, how might the pilot section flee the crew compartment?

BOB LAZAR'S SPORT MODEL

Despite Bob's protests against the limelight, his long-time friend, George Knapp, along with Jeremy Corbell, have been able to cajole him into making the rounds. Still, that doesn't make the man easy to reach.

As much as I'd like to do shots with Bob in some tavern, such opportunities seem few and far between.

BUT IF I COULD ASK...

1. Since you've been inside the craft and seen a test flight, did you gather anything about how its door opened and closed? I'd be curious to know how it seals shut. Not to mention, I'd like to know what mechanism would trigger it to open from the outside.
2. Have you thought about how it glowed so brilliantly during night flights but didn't have an overall glow during the daylight test?

3. Do you think the antenna/waveguide could be today's translucent aluminum? Oxford scientists manufactured it for the first time in 2009. Since it is optically transparent from near-ultraviolet to mid-infrared, might it work as a reflective waveguide in the microwave range?

WHAT SORT OF CRAFT IS THE SPORT MODEL?

Outside those questions, there are many more Bob wouldn't be able to answer. During interviews and podcasts, he shared the Sport Model had no provisions or, funny enough, bathroom facilities. Running under the assumption that all living organisms consume sustenance and produce waste, it would be a logical conclusion that this craft is essentially... a car.

It's not a sea-going vessel or a recreational vehicle as we would know. Those sorts of craft have places to store food and lavatories. This starship is barebones transport. Without knowing the occupant's physiology, an anthropomorphic guess would likely limit them to a few hours travel time. That leaves:

- a mothership
- an undersea base
- a base within our solar system
- distance to their home world is quickly traversable
- extra-dimensional rift
- time travel (they're from the future or past)
- Grays are androids or avatars (no digestive system)

HOW ARE THESE CRAFT CONTROLLED?

Bob also stated there were no obvious flight controls inside the craft. No throttles, wheels, buttons, levers, or dials. Thought steers the Sport Model. *Yeah, welcome to the woo.* Nearly every account relating to these beings characterizes them as telepathic. As much as we upright primates like to scoff, I've waded through countless witness testimonies that all dovetail back to extra-sensory perception (ESP).

Bob also related witnessing an S-4 daylight test flight. During that flight, a human operated the ship from within the craft. So, how was this done?

Apparently, this mental ability is something which can be, and has been, taught to select military pilots. Which leads me to wonder about those techniques. Were these skills developed under Hal Puthoff?

Are they similar to a USSR program under Lt. General Alexi Savin? Within Russian circles, he claims to have perfected remote viewing techniques. After teaching operatives, he recruited six average house-wives to show that anyone can achieve the same results. According to Savin, they turned out to be his most effective viewers.

SALVATORE PAIS' NAVY PATENT

I applaud Dr. Pais' devotion to his country. He offered his patents to the US Navy for free. This way industrial contractors could not charge exorbitant fees for their proprietary research, and the Navy Depart-ment could get bids for the project from companies.

How I wish we could meet for lunch, maybe over Maryland crab cakes and coffee. However, I'm sure he's a very busy man.

BUT IF I COULD ASK...

1. How does the crew see outside your craft?
2. Are we talking about external cameras and VR headsets?
3. If so, would the polarized vacuum disrupt the camera's optics?
4. If the craft's inertial mass is reduced, wouldn't it require greater gravitational waves for propulsion?
5. Why didn't the Navy classify your patents? The military has all the authority to do so if they wish.

Actually, Salvatore answers that last one. He feels, in the Navy's eyes, the best place to hide such an advanced invention is in plain sight. Coincidentally, in the book *In Plain Sight* by Ross Coulthart, the jour-nalist/author believes the patents may be a form of disinformation. Dr.

Pais flatly denies this. His work is legitimate. One should simply do the math.

I've given the disinformation possibility some thought. Perhaps there's an alternative Dr. Pais hasn't considered. What if your work is legit? Prototyping such a vehicle would require enormous sums of funding and engineers of high regard. It would take years of dedication and additional expenses to maintain secrecy. Such a project would likely only be attempted by one of the world's superpowers.

Now, here's a wild thought. What if the US already has a craft that runs circles around this invention? Wouldn't it be in their best interest to dangle these patents before Russia and China in hopes of getting them to focus resources on them?

Even if they could achieve an inertial mass reduction craft, it would be a prototype vehicle, likely with a lot of flaws; a Ford Motor Company quadricycle compared to a Chevy Corvette.

Food for thought.

WHY DOESN'T LIGO DETECT ALL THESE CRAFT?

I've seen a few science figures dismiss the notion of craft being propelled by gravitational drives as nonsense. Their argument is, "If all these spaceships are zipping around our planet, we could detect their gravity waves." They'll site LIGO (the Laser Interferometer Gravitational-Wave Observatory) as their source of proof. This is a knee-jerk reaction where little thought has been applied.

LIGO is a fascinating set of two observatories that use lasers and extremely accurate instruments to detect extraordinarily tiny variances in the laser's stretch as a gravity wave passes through the Earth. They can only use this in the rare instances when stellar objects collide with one another. The gravity waves these collisions release are of the lowest frequencies.

A high frequency gravity wave would be very local; short, tight waves that disperse in the vicinity of use. If you drop a stone in one end of the Great Lakes, do you expect to see the ripple on the other side?

JUST WHO *ARE* WE DEALING WITH?

That's the biggest question of all, right?

Our woo-train may very well fly off the rails at this point. If you were to ask people on the street, "Who are the UFO pilots?", most would say extraterrestrials. However, among ufologist, a broad consensus has expanded and is becoming widely embraced.

The very human want for a simple solution is almost never the case. Considering all the wild forms this phenomenon comes in — saucers, cigars, triangles, chevrons, wedges, orbs, plasma, etc. — we're looking at more than one non-human intelligence. If you care to include reports from Skinwalker Ranch, then we're also talking about expanding things to otherworldly wildlife too.

While beings from other planets appears to be a common go-to response, most scientists involved are expecting an equally diverse explanation of "who" they could be. They're leaving open notions like alternate dimensions and time travel, even alternate timelines.

Rest assured, when the day of disclosure comes, it's going to lead to a lot more questions.

CHAPTER 16
CONJECTURE

THINGS START to go somewhat radical with this chapter.

Up until now, we've presented several forms of UAPs and the physics encompassing their behavior. As compelling as all that is, one cannot help but wonder just what the hell is going on here. So, we're going to slide back a few steps from science basics and explore the hypothetical. I'll begin by pointing out what conjecture is.

A conjecture is a reasoned idea based on existing knowledge, open to verification, and subject to revision or rejection if new evidence emerges. It is rooted in logic, observation, and/or partial data. Often used in science, mathematics, and philosophy. The idea is to encourage skepticism and further inquiry. Essentially, it is a thought-experiment opened up for others to examine.

Throughout this guidebook, you've likely noticed a common thread. We started with a supposition, which is: governing powers within the US are withholding Earth-shaking technology and concealing wider universal societies.

But why would that be?

Obviously, the US military would want to hold hidden aces in their sleeves for tactical advantages later. And you can certainly see such good sense if they've recovered interstellar craft. It's a no-brainer the military would classify these things. Secret aces can only be played when the other players aren't aware they are present.

All such technology must go dark. The more powerful the tech, the farther it has to be distanced from oversight. That said, perhaps you've heard the maxim about paving roads to places with good intentions?

Over the course of collating info for this book, I've heard and read all sorts of wacky, woo-woo stuff. The vast majority were so far removed from my physics purview, there wasn't even a need to dismiss them. But along the way, a plausible narrative coalesced, an unacknowledged past that could explain most of what's at play.

CONSPIRACY TIME

Put on your tinfoil hats for this chapter, folks. History is about to get buck wild. Let's begin with a pivotal starting moment.

"Fatman" Atomic Bomb

In the 1942 Manhattan Project, the military brought together a group of PhDs — Enrico Fermi, John Dunning, Harold Urey, and Eugene Wigner — to see if an atomic bomb was possible. They quietly worked at Columbia University, New York. Edward Teller later joined them at the University of Chicago. As their research advanced, so did other secretive measures:

1. **Limited Access:** Only a small number of people had admittance to sensitive information. Security clearances were required, and those who worked on the project were carefully vetted.
2. **Compartmentalization:** The project was divided into many small parts, with each group working on a specific component. This prevented any one person from having a complete picture of the project.
3. **Restricted Communication:** Discussions between different parts of the project were limited. For example, mail was censored and phone calls were monitored to prevent information leaks.
4. **Disinformation:** False information was deliberately leaked to distract and mislead any potential spies.
5. **Secrecy Pledge:** Everyone was required to sign an oath, promising not to disclose any information about the project.
6. **Security Measures:** As nationwide locations were added, physical measures were implemented, such as armed guards, fences, and checkpoints to prevent unauthorized access to these expanded facilities.

Any of this ringing bells?

This led to the first atomic bomb detonating in July 1945 at Alamogordo, New Mexico. Weeks later, the A-bomb was dropped on Hiroshima and Nagasaki.

FIRST CONTACT

It seems those nuclear explosions sent out a signal to many other someones as well. The surrounding period witnessed an abundance of unidentified anomalous phenomena. One of the most famous of these incidents was a UFO crash and recovery in Roswell in July 1947.

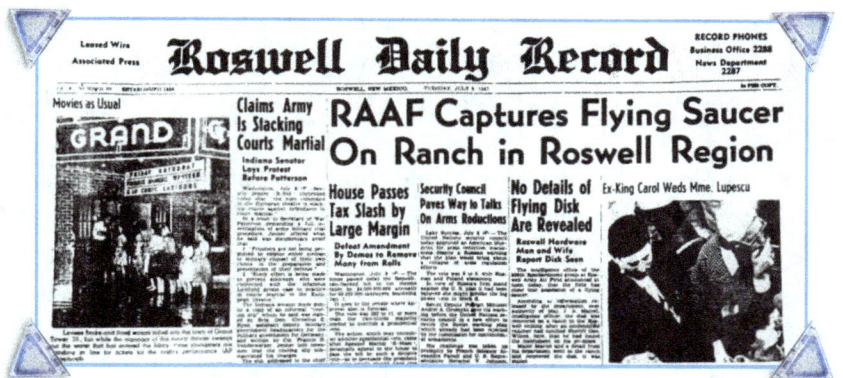

The accounts of Roswell vary greatly depending on who is delivering them. Yet, more than one account mentions bodies recovered and even a crash survivor. The craft occupants were your classic Gray aliens. Their bodies were preserved and stored, and the sole survivor, referred to as EBE (Extraterrestrial Biological Entity), was secreted away.

Over the next five years, communication was eventually achieved with EBE, and a channel of dialogue was established with other civilizations. Cruelly, accounts ended with EBE dying in 1952, well before being returned.

A couple of months after the crash (September 1947), two new federal agencies were founded: the United States Air Force and the Central Intelligence Agency. To this day, both agencies have remained tight-lipped on the UAP subject. Despite the Air Force clearly having a preponderance of evidence, it has been the US Navy which has proven to be the most forthcoming branch.

NEGOTIATIONS OPEN

On February 20–21, 1954, while golfing in Palm Springs, California, President Dwight D. Eisenhower received an outreach call. Another race of aliens was ready to meet. Under the guise of a dental emergency (a chipped crown), President Eisenhower flew in the dead of night to Edwards Air Force Base and met with a group of beings called the Pleiadians (Nordics).

The most bizarre trait about these beings is how nearly indistin-guishable they are from humans. They typically have Scandinavian features: six to seven feet (182–213 cm) tall, fair complexion, blond hair, blue eyes.

These Nordics focused upon the threat of nuclear war on Earth. They warned how such weapons are too disruptive to spacetime and their neighboring societies. They offered an exchange. The Nordics would share their advanced technology and philosophies if Eisen-hower would eliminate the nukes.

By this time, however, the Soviet Union had already developed their own A-bombs. The thought of giving up a US advantage was deemed out of the question. President Eisenhower declined.

NEGOTIATIONS ROUND TWO

A year later, February 11, 1955, President Eisenhower was documented as having "the sniffles" and was to spend time at rest after a morning of quail hunting. However, witnesses that morning reported Air Force One landing at Holloman Air Force Base, New Mexico. Several camera crews (two on the ground and one in a helicopter) were already on site with the expectation of a training documentary.

The encounter was said to have been filmed by this crew. At one point, the footage was to be disclosed in a television program, *UFOs: Past, Present, and Future*, hosted by Rod Serling. Just before going on air, the film was withdrawn. In its place, an eight-second clip was famously shown of a brilliant light lowering before the local mountain range. While many believe this to be from the original film, upon closer inspection, it turned out to be a jet fighter's headlight while on final approach.[1]

UFOs: Past Present and Future (1974) | *Sandler Institutional Films*

That morning's actual encounter began when three metallic saucers descended on the base. One of the crafts landed on the runway, and three human-sized beings stepped out. They were described as having blue/gray skin, a prominent nose (almost Middle Eastern-like), vertical-iris eyes, and wearing form-fitting jumpsuits. They also wore a headdress which acted as an interpreter. Met on the runway by two Air Force officers, they were escorted to the base facilities.

This summit concluded with a counter (more controversial) proposal. These other beings would share technology. In exchange, the US government would allow for animal experimentation and human studies, but only so long as the humans were returned unharmed.

But wait! There's more.

The aliens provided the locations of several tucked-away craft already on Earth. These would be excavated and collected at Area 51's S-4 facilities. A contingent of Grays would remain on the base for their own studies and to assist as guides. We humans still had to back-engineer everything, but they would give pointers along the way. This arrangement remained in effect until the Grays fled the base dramatically in the mid-seventies.

The last, most vital part of their agreement: No government is to openly acknowledge the existence of extraterrestrial life. So long as there is no disclosure, their agreement would stand.

Perhaps one day, once we've understood interstellar travel and elevated to a semblance of cultural maturity, Earth might be welcomed into their wider community. But it will be up to that community to decide when disclosure could happen, not an Earth government.

Until then, *aliens do not exist.*

GOING FORWARD

Okey dokey, folks. If we're to contemplate such a hypothesis, how might this conspiracy play out?

Despite how often pundits state that the government is leaky, that most often revolves around political agendas and cross-party sniping. When it's about National Security, discretion is paramount. Covert programs like the SR-71 Blackbird operated for decades, only coming to light just as it was being retired. By the time you saw stealth fighters and bombers in air shows, Aurora had already supplanted them. And you hardly know anything about Aurora today.

That said, the military would still have several problems on their hands. With US tax funding comes Congressional oversight. Politicians revolved in and out of office so often, it wasn't reasonable to believe they could remain tight-lipped about something so grandiose. On top of that, the 1967 Freedom of Information Act (FOIA) heightened the challenge.

How could they back-engineer recovered craft and do so without the revealing light of government oversight? The loophole became a chestnut in US governance.

They privatized it.

Once ownership was transferred to industrial military contractors, the secret was no longer in the possession of the US government. If a congressperson were to ask, the military will say they have no such technology. Thanks to "personhood" rulings, corporations are also afforded greater rights to privacy than even everyday citizens. No federal agency has access to non-funded proprietary development, which a corporation doesn't allow. The only exception would be if it fell under a criminal investigation.

With potential leaks shored up, what's to be done about the public fervor that sprung from so many sightings in the forties and fifties? That's where we enter Project Blue Book. While publicly promising to investigate the phenomena, their core role was to dismiss the topic. Internally, the team espoused the subject as a vulnerability Communist Russia could exploit to foster hysteria. Their strongest tool was to stigmatize those who took the subject seriously as simply loons.

And this has worked astoundingly well ever since.

With everyday citizens placated, other challenges remained. How do contractors study the subject while hiding their advancements? They would need great minds and a way to keep the rest of science at bay. If only they could offer a misdirection, some theory that couldn't be substantiated.

TWO WORLDS OF SCIENCE

Between 1968 and 1970, string theory came into being. By the eighties, this became the go-to place for young physicists to gather. And for the decades since, string theory, and all its diverse interpretations, has been a grand thought experiment, a complex Rube Goldberg device achieving next to nothing. Outside of some mathematical formulas used to understand black holes and condensed matter, the theory hasn't moved the physics needle any further than where it was left in the 1940s.

There has been no new physics.

The world has certainly become a technological marvel. From transistors to microprocessors and superconductors, all those advancements were built upon quantum mechanics. The Large Hadron Collider (LHC) confirmed a whole zoo of baryonic particles. But outside of stylistic or ergonomic comforts, what makes our lives any different from those living in the forties or fifties?

Recently, Harvard mathematical physicist Eric Weinstein put forth a thought-provoking observation. In 1953, two successful entrepreneurial figures, Agnew Bahnson and Roger Babson, offered funding to research anti-gravity. Bahnson and Babson formed a collective to find a way to defeat gravity, called the Gravity Research Foundation. Minds like Dr. Bryce DeWitt and Dr. Louis Witten (father of string theorist Edward Witten) headed up this organization.

Initial excitement brewed around the possibilities before fading into the background. One could surmise their research either fizzled out... or possibly sparked something new, then went black.

ANOTHER MANHATTAN PROJECT?

As secretive as the original A-bomb project was, it could have been noticed had someone been paying attention to the particular field of PhDs being gathered.

For example, a purview of the National Security Agency (NSA) revolves around decoding cryptographic messages. When you look at all the number-theory PhDs that specialize in such things, you'll notice many top graduates don't end up in private or collegiate sectors. Though seemingly unemployed, many of these people now, coincidently, live in the same area of Virginia.

Let's say you want a UFO/UAP Manhattan Project to suss out gravity, while also flying under the public radar. Eric Weinstein suggested the fields that would be required. You'd need PhDs in high-energy (particle) physics, differential geometry, and general relativity. Where have these best and brightest minds ended up today?

The first is the University of Texas at Austin's Gravitational Physics Department. However, even more notable is the State University of Stony Brook, Long Island. Within those halls are an astounding collection of minds at a very average State College setting. Yet, they're not trumpeting horns or selling the University as the powerhouse it is.

Next, how might we fund this research?

Many don't realize mathematics and physics PhDs are also some of the most sought-after people at financial hedge funds. Their algorithmic expertise has frequently proven to outpace markets.

Over the last ten years, there have been four hedge funds that appeared to be reaping in inexplicable financial gains: Bernie Madoff, Jeffrey Epstein, D.E. Shaw, and Renaissance Technologies.

Bernie Madoff (imprisoned) was revealed to be a Ponzi-schemer. The funding source of Jeffrey Epstein (deceased) has not yet been revealed and is unlikely to ever be. The other two funds operated with a focus on quantitative analysis and technology. According to the late Berkley emeritus mathematician, Isadore Singer, the world's greatest mathematics and physics department is Renaissance Technologies.

Stony Brook University

Renaissance

Both Stony Brook U. and Renaissance Technologies are located in the same small town of East Setauket, New York.

PREPARING FOR DISCLOSURE

Since we're playing with conjecture, how about some fun stuff?

If you knew there'd be many decades before a day of disclosure would come, how might that transition be best smoothed over? One solution would be to insert some useful concepts into popular culture.

As a kid, I used to scoff at ridiculous movie and television sci-fi aliens like those in *Star Trek* and *Star Wars*. Anyone with a half a brain knows life out there would look nothing like us humans. The genetic odds would be obscenely beyond the pale. Sure, there are evolutionary benefits to a bipedal form, but not every world is going to have the exact conditions our ancestors adapted to. An extraterrestrial would never look human.

Yet, there have been these encounters with human-like Pleiadians. Most notable of these were with the abduction/rescue of Travis Walton, a UK encounter with Jessie Roestenberg, and a long-term relationship with Swiss eccentric, Billy Meier. As kooky at Billy is, even Bob Lazar noted the photos he took of Pleiadian craft were identical to some of the nine he witnessed at S4.

While still in my youth, I found Gene Roddenberry's *Star Trek* concept of a Prime Directive to be prescient. His fictional Federation of Planets had an oath of non-interference with primitive worlds. Even if Captain Kirk dramatically circumvented them, the overall ideal seemed a sound one. A requirement of alien nondisclosure would fall hand in hand with such a premise.

It is also rumored that upon seeing Stephen Spielberg's *Close Encounters of the Third Kind* at the White House, Ronald Reagan quipped, "You don't know how true this really is."

Which brings us back around to the current day and DoD disclosures of UAPs.

JOHN RAMIREZ, CIA OFFICER (RET.)

In recent interviews, a retired CIA officer opened up about goings on behind the scenes. John Ramirez served within the agency from 1984 to 2009. He held the rank of GS-15, which is a senior-level position within the agency. During his tenure, John worked in various departments, including the Directorate of Science and Technology, the Directorate of Intelligence, and the Office of the Director of National Intelligence (ODNI).

John also straddled a unique position from within. He sat squarely on a fence between the agency and his personal fascination with UFOs. He'd often spend his vacations attending MUFON conventions and engaging in minor speaking events. All the while, he openly disclosed his extra-curricular activities to his agency superiors.

While tenured within the agency, he was even offered a chance to be "read in" on some UAP projects. He declined, knowing he'd have to give up his ability to speak on the topic. Over the course of his twenty-five years moving within sensitive circles, there were many things he picked up either indirectly or unintentionally. But since he never signed a UAP non-disclosure agreement (NDA), they cannot prosecute him for sharing what he learned.

John continues to follow current briefings and ongoings with the All-domain Anomaly Resolution Office (AARO), as that agency is required to debrief Congress of its findings.

We're all familiar with how politicians two-step around the truth or feint on promises without ever landing them. Federal office holders also dance with the truth, but in specific ways. John suggests paying close attention to how answers are parsed. They are well practiced at answering only the question as presented.

Be direct. Keep answers short. Never offer more than what was asked. Understand that some details are still shielded as classified. If ever cornered without a way to obfuscate or redirect, simply lean on ignorance.

More importantly, be sure whatever role you've been assigned is curtailed within tight parameters. As an example, contrary to popular thinking, AARO is not a disclosure agency. Their domain is in regard to national air safety. Examinations are solely about craft sightings and a threat of midair collisions. If you get questions about what happened at Roswell or whether someone is holding off-Earth materials, well, that isn't in their purview. Also, AARO isn't tasked with actively tracking down leads; they are only to respond to what they passively receive.

Today, we appear to be drawing closer to a "day of disclosure." Yet, the US government seems to be holding tightly to some agreed upon Prime Directive. They can now discuss unknown objects in space, the sky, and the sea, but draw a clear line with whoever occupies these vehicles.

There is an overall concern that if other civilizations were to reveal themselves, it could lead to a mass panic.

COUNTDOWN TO DISCLOSURE

John Ramirez's most compelling statement was how close that day might very well be. During a *Podcast UFO* interview, John offered his belief that the 2017 *New York Times* article marked the beginning of a countdown. It is why representatives from AATIP have entered mainstream media to open up about the topic.

> "I think the word got out within the government that they're showing up in 2027. And we better be prepared. If not, there's going to be a lot of explaining to do. So, I think that dialogue has happened within certain areas inside the government, that we need to prepare.
>
> "That's why 2017 set a clock of ten years, and why Luis Elizondo earlier in this year said, 'Just find a hobby for five years, and it'll all be out,' and he said that this year [2022].
>
> "I have heard 2027 in an official capacity, which I can't reveal.* So, I would say that people in the government are aware of something happening, and there's limited time, a few more years to prepare the people. That is what's wrapping up this acceleration from the previous seven decades of not even acknowledging it, to now acknowledging it at a faster and faster pace."

*John later shared the official capacity he referred to. It took place in a CIA Secure Compartmented Information Facility (SCIF). A SCIF is a highly secure area designed to safeguard sensitive and classified information from unauthorized access or interception. These room-sized, spy-proof boxes have been popularized in many dramatized thrillers.

As one meeting ended within the SCIF, other officers present began speaking on the 2027 topic. That was before being corrected about how John had not been read in.

2025 Update: There has been some discussion in the UFO community that Mr. Ramirez's conjecture could be attributed to a psychological operation. That this is a build up to enhance a potential false threat narrative. Like all things in this guide, weigh for yourself how best to consider the data.

Speaking of that narrative…

CHAPTER 17
THREAT DIVISION

ASK YOURSELF, how is our world contending with this new UAP paradigm shift?

The vast majority (those who wouldn't pick up a book like this) dismisses it off-hand. Life goes on, with bills to pay and better things to attend to. No serious person gives it more than a second thought thanks to a well-entrenched stigma.

For others, well…

Where you learn about the phenomena determines what sort of slant you receive. Like many overarching concerns, facts can be presented one way or another based on who is driving the discussion. As of this publication, that discourse has fallen within two camps: one, where the UAP poses a threat, and the other where the UAPs are benevolent watchers.

NON-HUMANS AS A THREAT

The US Department of Defense, cable news, and print media outlets mostly run on the old maxim, if it bleeds, it leads. Tension, fear, concern — these traits command attention. Nothing turns heads or funds military projects better than a good threat.

Spokespersons presenting from this angle are quick to point out, a threat doesn't mean the same thing as intent. If hobbyist drone pilots are bopping around an airport operation area, they are a threat to aviation safety regardless of whether their reasons are ignorance or stupidity. Even flocks of birds are considered threats for this same reason.

Luis Elizondo and Christopher Mellon

Thanks to Luis Elizondo, Christopher Mellon, and To the Stars Academy; they've declassified several cases of UAP impinging on Naval training ranges. When you have fighter jets ripping through the air at high speed, an unexpected craft appearing is a serious hazard. That alone makes them a threat.

But take it a step further. Why are they there? Is it to gather intelligence on human military capabilities?

There have even been reports of UAPs interfering with nuclear silos. In 1967, at Malmstrom Air Force Base, a glowing red orb hovered over their flight control center. Within minutes, it shut down ten intercontinental ballistic missiles (ICBMs). Later, in 1982, at a Russian missile base in Ukraine, a disc-shaped UFO appeared and activated their ICBMs. Their nukes began preparing for launch, only to mysteriously wind back down after fifteen heart-pounding seconds.

Who wouldn't consider toying with the possibility of World War III a threat?

Then you also have tens of thousands of cattle mutilations reported worldwide. And not just cattle — bodies with these distinctive characteristics have included sheep, horses, goats, pigs, rabbits, cats, dogs, bison, deer, and elk. These often coincide with UAP sightings, and inexplicable animal-recovery scenes:

1. **Exsanguination:** Remains are fully drained of blood, leaving none within the body or in the surroundings.
2. **Organ Removal:** Surgically extracted with scalpel precision, body parts — genitals, eyes, ears, tongue, and lip regions down to the jawbone — are typically missing.
3. **Digestive End Coring:** A cylindrical extraction of the rectal region is commonplace.
4. No signs of restraints or struggles

Even worse, there are reasons to believe these mutilations are not limited to livestock. There is the 1956 case of USAF Sgt. Jonathan P. Lovett in New Mexico. And in 1982, there was a body recovery involving a middle-aged male from a reservoir in Sao Paulo, Brazil. Or even recently (2023), attacks were reported by Peruvian villagers of seven foot tall aliens they call *pelacaras* (face peelers). These incidents were identical to those with the cattle.

As disturbing as these reports are, you can easily see why they hold sway in several important circles.

I'm certainly not casting aspersions at those working under a threat assessment. If I were within the government and needed funding for research, there is an understanding of what motivates best. When what you have to work with is a hammer, all your problems look like nails.

NON-HUMANS AS BENEVOLENT

If you're looking for the counterargument to the threat aspect, you'll inevitably come upon another figure in the UFO community. Dr. Steven Greer is a retired emergency room surgeon turned founder of the Center for the Study of Extraterrestrial Intelligence (CSETI) and the Disclosure

Project. He's produced several documentaries on the subject; many can be found on streaming services and YouTube.

Right off the bat, I had reservations about Dr. Greer. He presents as your classic woo-woo spiritual leader, looking to expand his flock. And even today, I still find it difficult to get past this.

Then I learned of his involvement in a very credible revelation.

THE WILSON/DAVIS MEMO

After the passing of Apollo astronaut, Edgar Mitchell, in 2016, a memorandum was found within his estate. It was of a 2002 beat-by-beat conversation between Admiral Thomas R. Wilson and a representative from a top-three defense contractor. Dr. Eric W. Davis was present as per the Admiral's request.

At that time, Admiral Wilson was the director of the Defense Intelligence Agency (J2), giving him full need-to-know access as a member of the President's Joint Chiefs. Dr. Davis is a physicist with the Institute for Advanced Studies in Austin, Texas. It was actually Eric Davis who documented this clandestine meeting.

The admiral had done his research in advance. He'd discovered several Unacknowledged Special Access Programs (uSAPs) dealing with off-world technology being kept from his office. After tracking down a few contacts, he flew to Las Vegas and met with a one of these representatives. This was a nearly two-hour meeting held in a parked car. The admiral arrived with the expectation of full disclosure for his department.

He failed to be read in.

The contractor's only reason to have the meeting was to discover who'd leaked their program's existence to the Joint Chiefs. That's when I was startled to learn of Dr. Greer's involvement.

Five years earlier, in 1997, Admiral Wilson had a summit with Dr. Greer, astronaut Edward Mitchell, and Lt. Cmdr. Willard Miller. They'd convened to inform Admiral Wilson of what was being withheld from his department. Greer provided the admiral a list of covert program code names and individuals his organization had uncovered, leading to these Unacknowledged SAPs. The admiral spent the intervening years scrubbing through restricted documents and connecting the dots Greer's team had provided.

Multiple reliable sources have confirmed and corroborated the Wilson/Davis memo. This clandestine meeting really happened. And, I invite you to read the fifteen-page document[1], which shares so much more.

SIRIUS DISCLOSURE'S SUPPOSITION

In the previous chapter, I put forth the sensible reaction of the military to classify retrieved craft as top secret. If Dr. Greer's understanding holds true, the original covert nature has been twisted into a historic level of corruption.

Once private industrialists saw successes in their back-engineering attempts, a greater understanding coalesced. If zero-point energy could be as ubiquitous and efficient as predicted, it would immediately become a threat to the petroleum industry and the United States as the world's reigning superpower.

With all the trillions of dollars spent worldwide on fossil-fuel energy, it would be easy to get industry powers to want to suppress the free energy aspect of the technology. Getting US government support only required explaining how America's sway in the world economy was deeply intertwined with the petrol economy.

If every country, big and small, had unlimited electricity that was non-polluting, and cost next to nothing to produce, over the course of a few decades, economies would level out. The US standing as a dominant force would be vastly diminished.

This would be an easy sell to bury the technology.

A DANGEROUS STUDY

Over several decades, many inventors that have worked on zero-point energy have met unceremonious ends.

1. **Rory Johnson** (Cold Fusion Vehicle Motor): 1979: The US Department of Energy placed a Secrecy Act of 1952 gag

order on his technology. Rory took all his inventions to California, hoping to evade restrictions. He died unexpectedly soon after of cardiac arrest.

2. **Stan Meyers** (Vehicle Engine run on Water): 1997: Died from a cerebral hemorrhage while having lunch with two investors. His last words were, "I was poisoned."

3. **Aire deGaus** (patented ZPE tech): 2007: On his way to meet with investors, they found him dead of cardiac arrest in an airport garage.

4. **Mark Tomion** (patented the ZPE Stardrive): 2009: After developing a prototype, he soon passed from a cardiac event. His research is now missing.

5. **Dimitry Petronov** (plasma battery inventor): 2010: After leaving a Moscow café, he was never seen again.

6. **James Allen** (documentary filmmaker on ZPE): 2013: Died within three months of a cancer diagnosis. His form of cancer typically takes years to develop, not weeks. An autopsy revealed James' blood contained twelve heavy metals and radioisotopes at toxic levels.

WHY ALL THE CRASHES?

This is a question I heard bandied about by both ufologists and debunkers. And it's a reasonable one. "How is it these ETs can build these spacecraft and cross countless light years all the way here, only to just take a nosedive and crash once they arrive? That makes no sense."

I agree. And it's not just a few. There's likely been a dozen or more of these crashes reported worldwide:

1. (1945) **The Trinity Incident:** New Mexico, near where the first atomic bomb was tested. According to accounts, a UFO crashed, and the wreckage was transported to the nearby Los Alamos National Laboratory.

2. (1947) **The Roswell Incident:** July crash on a ranch near

Roswell, New Mexico. The US military later claimed it was a weather balloon.

3. (1957) **The Ubatuba Incident:** On a remote beach in Brazil, witnesses observed a speeding object descend from the sky. Before it could strike the sea, it redirected upward, then exploded, throwing debris into the water and along the beach. Locals collected pieces and sent them off for study. While isotopic ratios mostly of magnesium were within terrestrial norms, the oddly present strontium would require artificial manufacturing.

4. (1967) **The Shag Harbour Incident:** In October, a crash happened in the water off Shag Harbour, Nova Scotia, Canada. Witnesses claimed to have seen a bright object crash into the water, and the Canadian military conducted a search and rescue operation.

5. (1986) **The Dalnegorsk Incident:** In January, a reported crash in Dalnegorsk, Russia. Witnesses saw a bright object crash into a mountain and found strange metal fragments at the crash site.

6. (1996) **The Varginha Incident:** In January, a crash occurred in Varginha, Brazil. Witnesses claimed to have seen strange creatures taken away by Brazilian military personnel. All debris and remains were turned over to the US Air Force.

John Ramirez offered one plausible explanation. What if these non-human cultures are barred from sharing their technology so instead "crash" their craft for us? This idea, taken from a retired CIA officer, would be the sort of useful loophole covert operators might employ.

However, Dr. Greer's explanation isn't nearly as reassuring. His opinion: military forces worldwide have been shooting them down, using classified high-energy weapons.

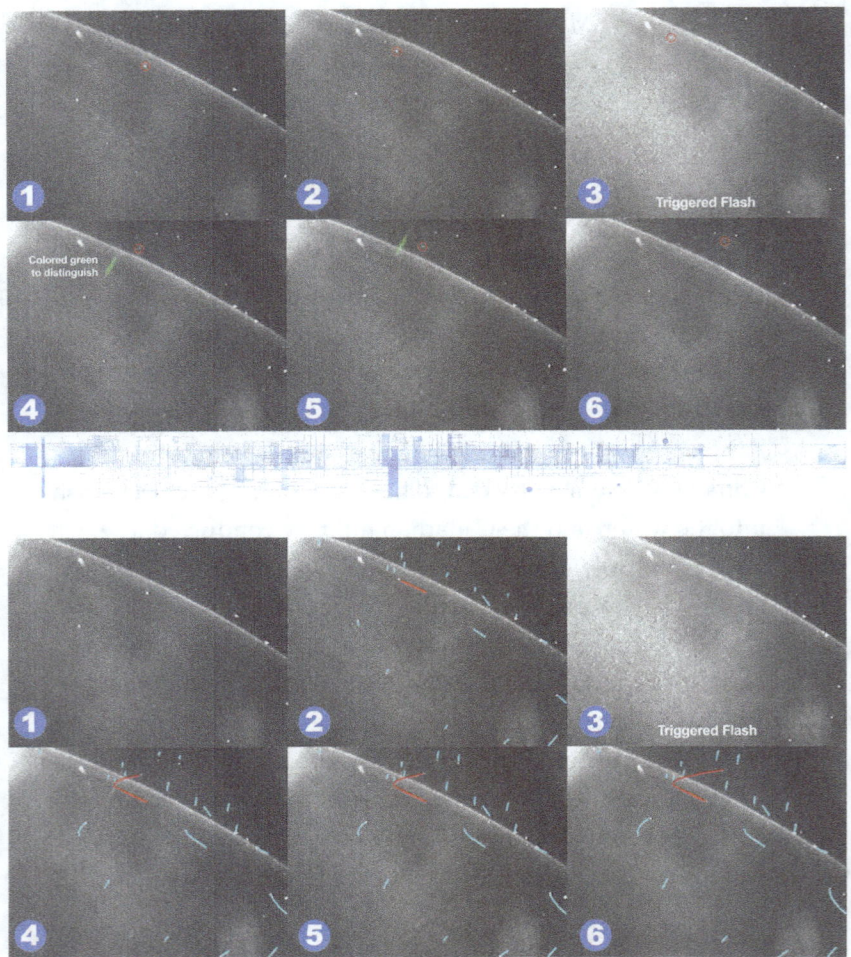

These frame-by-frame images are from a 1991 Space Shuttle *Discovery* video. At the moment a flash is recorded, the object tracking into the upper left abruptly banks. It accelerates away, heading out into space. Just after, an Earth-based streak crosses the object's original course. Debunkers cite this as just ice crystals from *Discovery*'s rockets. Yet, look for yourself at how this one object tracks differently from the others.

If what Dr. Greer is espousing turns out to be true, then it is us humans who are the real threat. His understanding involves private contractors forming a cabal in the United States. This cabal has next-level technology: its own military units, the ability to steer current US governance, and enough deep funding to direct the pending era ahead.

> **Senator Daniel K. Inouye:** "There exists a 'shadowy government' with its own Air Force, its own Navy, its own fundraising mechanism, and the ability to pursue its own ideas of national interest free from all checks and balances, and free from the law itself."

> **President Bill Clinton:** During an interview with long-time White House reporter, Sarah McClendon. "Sarah, there's a government inside the government, and I don't control it."

Dr. Greer is warning of a "false flag" operation in the works. They would fake an alien invasion, using their back-engineered technology. The conspiracy would require scaring the world enough for a restructure under a new unified government, one of this cabal's design. That way, humanity would turn away from outside societies, control of the populace could be maintained, and the rest of the universe would have to negotiate with their cabal leadership.

> **President Dwight Eisenhower:** "In the councils of government, we must guard against the acquisition of unwarranted influence, whether sought or unsought, by the military-industrial complex. The potential for the disastrous rise of misplaced power exists and will persist. We must never let the weight of this combination endanger our liberties or democratic processes. We should take nothing for granted. Only an alert and knowledgeable citizenry can compel the proper meshing of the huge industrial and military machinery of defense with our peaceful methods and goals, so that security and liberty may prosper together."

THREAT OR BENEVOLENCE: WHICH IS RIGHT?

There is compelling evidence coming out of both sides of this argument. Of course, I'd like to believe former cavemen with iPhones could get along with those more evolved.

Here's food for thought: Think about the incredible technology presented in this book. I'll even posit that a craft like Lazar's Sport Model might only be a single rung up on a scientific ladder from where we are now. Yet, if aliens are here with such advanced power, they could've easily overrun our society and wiped us out at any point.

But they haven't.

Martin Luther King once paraphrased, "The arc of the moral universe is long, but it bends toward justice." What if we play out that ideal over hundreds, thousands, even a million years? If other entities outside Earth are so far ahead of us technologically, might they be philosophically as well?

There is a counter argument/expression, however:

The more things change, the more they stay the same. That implies, despite all our advancements and life improvements, our societal vulnerabilities and pitfalls are really not much different from where they were a hundred or a thousand years ago. Humans are still flawed, territorial, and violent at times. We evolved from predators to become the dominant species on our world. Should we believe non-humans are any different?

In the end, I honestly don't know.

Is there an empirical way we can determine one way or another, or possibly mix both? I'd guess the truth is likely somewhere in the middle of these factions.

If we cannot decide, our inaction will render the point moot. Future courses are set by those willing and able to navigate. And at the moment, it appears we, the people of Earth, are *not* at the helm.

CHAPTER 18
SUMMATION

LET'S take a breath and tip our tin hats back.

Most everyday people will take the last two chapters as straight-up conspiracy theory. To that, I will counter with my choice of "Conjecture" as a chapter title.

The point of conjecture is to engage in speculative reasoning or educated guesses about a particular topic or problem when there is limited or incomplete information available. It allows reasonable people to explore possibilities, formulate hypotheses, and generate potential explanations or solutions.

It forms a starting point for further exploration and refinement of ideas. That's where you should bring to bear your critical thinking. It is through conjecture we examine the evidence, and test hypotheses.

Don't think my conjecture holds up?

Show me where. Who knows? I may very well agree with you. Many of my assertions are likely not going to pan out under scrutiny. I'm often wrong. But that's all right. We should be willing and open to finding the flaws in what we put forth. Only then can we improve our understanding. *That is what science does...* how we advance ourselves as a whole.

Dismissing it as fringe is ending the examination before it begins.

Fringe may not be at the heart of stigma, but it has been employed as a wedge—a way to drive great minds away from examining. It has split the field into debunkers or crackpots. And in this modern but divided world, who really wants to be lumped on either side?

In 1905, Albert Einstein presented special relativity to the scientific community. While there was some initial push back, it was short-lived. Within a brief span, it was scrutinized and accepted into the greater whole. I wonder, if he were to make such assertions today, would anyone listen?

With this guidebook into the world of physics, I realize now how I entered with raised expectations. I envisioned great minds of our time as rising above the mistakes of the past. They'd be objective thinkers, willing and open to discuss ideas. They would strip emotion from their equations while working toward finding new enlightened paths to truth.

Oh... That did not turn out to be the case.

It would seem the religious zealotry of old is just as comfortable within academia. Rather than ardent worship to one god or another, they are now ensconced churches to established scientific dogma. Many scientists take strong positions that their own theories are the only true way, and others are just flat wrong. My attempts to under-stand various physics equations and concepts when posted to physics forums were promptly blocked and penalized.

Would Einstein even be allowed a seat alongside Newton today?

That said, we are not too far gone. I applaud voices like Eric Wein-stein, Sabine Hossenfelder, Michio Kaku, and Avi Loeb as they break from convention and explore novel ideas. I believe we need more wild thinkers willing and able to get into the scrum.

THE WAY FORWARD

As of this writing, there have been Congressional hearings on UAPs. The whistleblower, David Grusch, and both retired Navy pilots, David Fravor and Ryan Graves, have entered sworn testimony into public record. Because of Mr. Grusch's standing in the intelligence commu-nity and his steadfast restraint on classified information, they have not been able to smear or prosecute him.

Waiting in the wings are another forty known whistleblowers and hundreds if not thousands more that are unknown. The difference between these whistleblowers and David is the first-hand testimonies they offer. These are the actual scientists, military personnel, federal agents, and private contractors who've worked directly in recovery and back-engineering projects. They have been waiting to see the path Mr. Grusch navigates through a minefield... and noting the markers placed along the way.

In an unprecedented move, spearheaded by Majority Leader Chuck Schumer, the Senate has presented an amendment to the annual defense spending bill. Among the components of the new legislation is an inclusion that the US federal government will have "eminent domain over any and all recovered technologies of unknown origin (TUO) and biological evidence of non-human intelligence (NHI) that may be controlled by private persons or entities in the interests of the public good."

Drawing additionally from legislation dedicated to uncovering the truth behind President John F. Kennedy's assassination, this proposal also orders executive branch agencies to relinquish UAP records to a review board. This board will operate under the guiding principle of "immediate disclosure." Rather than hiding behind bureaucracy, these agencies will be forced to either disclose or justify why a document should remain classified.

Either way, someone will have to confess to having something.

Admittedly, I have my doubts it will survive in its current form. While this amendment has teeth, it is about to be sent through a dental gauntlet. None of it will matter if it isn't signed into law. That doesn't happen unless President Biden puts pen to paper. He is about to enter his re-election campaign.

But who knows?

Can you think of a better way to have a lasting legacy? It'll be on him to decide if he is that man... Or, will he leave it for whoever comes next?

ON A POSITIVE NOTE

One widely accepted reason for *Star Trek*'s popularity was how Gene Roddenberry had crafted a more hopeful tomorrow. It wasn't some dystopian, post-apocalyptic irradiated future (the sort we appear to be politically tipping toward at the moment). Gene's Federation was an explorative society, eager to seek out new life and new civilizations.

I'm honestly excited for what the future may hold. To be here, when our civilization is on the verge of crossing one of the most historic thresholds possible, I can't help but feel fortunate.

And I can't wait to see what our children may yet see.

DON'T STOP NOW

This is only the first in our series. If this has sparked your curiosity and you're hungry for more, I can offer up something from the cereal food group in our next inquiry. Crop Circle Casework explores our wild world while literally out in the field.

If like myself, you've seen all those stunning photos of complex crop circles, you're likely questioning how such phenomena can be possible. Believe it or not, the next book does have those answers and so much more.

Jump back over to our retailer and pick up your next guidebook.

WOOF! THAT WAS A LOT TO DIGEST.

NOW THAT YOU'VE reached the end, what did ya think?

This is my first ever UFO Science guidebook. If you'd like me to continue with this as part of a series, I really could use your feedback.

A quickie 2-minute post via the review links below is what I need more than anything. Each review boosts our little UFO book upward in the ranks. Honest reviews not only give me the warm fuzzies (most often) but also help online retailers to distinguish exceptional books. That in turn means they'll show it to more readers.

If you'd like to see that happen…

CLICK BELOW TO REVIEW ON
GoodReads

ACRONYMS | GLOSSARY

- AARO: **All-domain Anomaly Resolution Office** - is an office within the United States Office of the Secretary of Defense that investigates unidentified flying objects (UFOs) and other phenomena in the air, sea, and/or space and/or on land: sometimes referred to as "unidentified aerial phenomena" or "unidentified anomalous phenomena" (UAP).
- AFB: **Air Force Base**
- ARV: **Alien Reproduction Vehicle**
- ATFLIR: **Advanced Targeting Forward-Looking Infrared** - AN/ASQ-228 is a multi-sensor, electro-optical targeting pod incorporating thermographic camera, low-light television camera, target laser rangefinder / laser designator, and laser spot tracker developed and manufactured by Raytheon.
- B.Eng: **Bachelors of Engineering**
- BBE: **Biefeld-Brown Effect** - an electrical phenomenon where high voltage/low amperage is applied to an asymmetric capacitor generating a net propulsive force toward the smaller electrode.
- CAP: **Combat Air Patrol** - provided over an objective area, the force protected, the critical area of a combat zone, or in an air defense area, for the purpose of intercepting and destroying hostile aircraft before they reach their targets.
- CAS: **Control Augmentation System** - is utilized to shape aircraft response to pilot inputs, as well as provide three-axis damping and autopilot functions.
- CIA: **Central Intelligence Agency**
- CSETI: **Center for the Study of Extraterrestrial Intelligence** - is an international nonprofit scientific research and education organization founded by Dr. Steven M. Greer and dedicated to the furtherance of our understanding of extraterrestrial intelligence.
- EBE: **Extraterrestrial Biological Entity** - the first alien captured from the UFO Roswell crash in 1947.
- EM: **Electromagnetic**
- ESP: **Extra Sensory Perception** - a paranormal ability pertaining to reception of information not gained through the recognized physical senses, but sensed with the mind.
- EED: **Extended Electrodynamics** - a modification of classical Maxwellian electrodynamics to restore scalar field outcomes and associated effects, such as a polarizable vacuum and higher-order field interactions.
- FBW: **Fly by Wire** - is a system that replaces the conventional manual flight controls of an aircraft with an electronic interface.

- FLIR: **Forward Looking Infrared** - technology which creates an infrared image of a scene without having to "scan" the scene with a moving sensor.
- FoIA: **Freedom of Information Act** - provides the public the right to request access to records from any federal agency.
- GHz: **Gigahertz** - a unit of frequency equal to one billion hertz
- GR: **General Relativity** - is the geometric theory of gravitation published by Albert Einstein in 1915 and is the current description of gravitation in modern physics.
- Hz: **Hertz** - the standard unit of frequency in the International System of Units, equal to one cycle per second.
- ICBM: **Intercontinental Ballistic Missile** - any supersonic missile that has a range of at least 3,500 nautical miles (6,500 km) and follows a ballistic trajectory after a powered, guided launching.
- ICIG: **Intelligence Community Inspector General** - conducts independent and objective audits, investigations, inspections, and reviews to promote economy, efficiency, effectiveness, and integration across the Intelligence Community.
- J2: **Intelligence (Joint Staff Directorate)** - supports the Chairman of the Joint Chiefs of Staff, the Secretary of Defense, Joint Staff and Unified Commands. It is the national level focal point for crisis intelligence support to military operations, indications and warning intelligence in DoD, and Unified Command intelligence requirements.
- LED: **Light-Emitting Diode** - a semiconductor device that emits light when current flows through it. Electrons in the semiconductor recombine with electron holes, releasing energy in the form of photons.
- LHC: **Large Hadron Collider** - is the world's largest and highest-energy particle collider. It was built by the European Organization for Nuclear Research between 1998 and 2008 in collaboration with over 10,000 scientists and hundreds of universities and laboratories, as well as more than 100 countries.
- LIGO: **Laser Interferometer Gravitational-Wave Observatory** - is a large-scale physics experiment and observatory designed to detect cosmic gravitational waves and to develop gravitational-wave observations as an astronomical tool.
- MUFON: **Mutual UFO Network** - is a US-based non-profit organization composed of civilian volunteers who study reported UFO sightings.
- NDA: **Non-Disclosure Agreement** - is a legally binding contract that establishes a confidential relationship.
- NGA: **National Geospatial-Intelligence Agency** - is a combat support agency within the United States Department of Defense whose primary mission is collecting, analyzing, and distributing geospatial intelligence in support of national security.
- NHI: **Non-Human Intelligence**
- NICAP: **National Investigations Committee On Aerial Phenomena** - an unidentified flying object research group most active in the United States

from the 1950s through the 1980s. Its current use is as a repository on UFO cases.

- NRO: **National Reconnaissance Office** - is a member of the United States Intelligence Community and an agency of the United States Department of Defense which designs, builds, launches, and operates the reconnaissance satellites of the U.S.
- NSA: **National Security Agency** - is responsible for global monitoring, collection, and processing of information and data for foreign and domestic intelligence and counterintelligence purposes, specializing in a discipline known as signals intelligence.
- ODNI: **Office of the Director of National Intelligence** - is a senior, cabinet-level United States government official, required by the Intelligence Reform and Terrorism Prevention Act of 2004 to serve as executive head of the United States Intelligence Community and to direct and oversee the National Intelligence Program.
- OAM: **Orbital Angular Momentum** - is the component of angular momentum of a light beam that is dependent on the field spatial distribution, and not on the polarization.
- PhD: **Philosophiae Doctor** - is the most common degree at the highest academic level, awarded following a course of study and research. PhDs are awarded for programs across the whole breadth of academic fields.
- QED: **Quantum Electrodynamics** - it describes how light and matter interacts, and is the first theory where full agreement between quantum mechanics and special relativity is achieved.
- QM: **Quantum Mechanics** - is a fundamental theory in physics that provides a description of the physical properties of nature at the scale of atoms and subatomic particles.
- RF: **Radio Frequency**
- Scalar Field: a mathematical function that assigns a value (number) to every point in spacetime. It is independent of direction and is often used in physics to describe temperature, gravitational potential, or quantum mediums such as the Higgs field.
- SCIF: **Sensitive Compartmented Information Facility** - is a secured and enclosed area within a building that is used to process sensitive compartmented information.
- SSP: **Strategic Systems Programs** - is responsible for managing the development, deployment, and maintenance of submarine-launched ballistic missile systems, playing a crucial role in ensuring the nation's sea-based nuclear deterrence capability.
- TUO: **Technologies of Unknown Origin**
- USAP: **Unacknowledged Special Access Program** - are security protocols made known only to authorized persons, including members of the appropriate committees of the United States Congress.
- UAP: **Unidentified Anomalous Phenomena** - observations that cannot be identified as aircraft or known natural phenomena from a scientific perspective.

- UAPTF: **Unidentified Aerial Phenomena Task Force** - a program within the United States Office of Naval Intelligence used to "standardize collection and reporting" of sightings of unidentified aerial phenomena.
- UFO: **Unidentified Flying Object** - any perceived aerial phenomenon that cannot be immediately identified or explained.
- VTOL: **Vertical Take-Off & Landing**
- WSO: **Weapon Systems Officer** - nicknamed "Wizzo", is an air flight officer directly involved in all air operations and weapon systems of a military aircraft.
- ZPE: **Zero-Point Energy** - is the lowest possible energy that a quantum mechanical system may have.

ABOUT THE AUTHOR

Who, me?

You might expect to learn of my advanced degrees in theoretical physics. Or how I'm an expert in geospatial relativity and high-energy particles here. I could present my honorary tenure with European Organization for Nuclear Research (CERN). As long as I did all of that with my tongue in cheek, you'd likely realize I was having my way with word-smithing the fantastic.

You see, I'm a novelist.

Rather than having an academic history in the Sciences, I started with a graphic design degree. Yeah. That was me doodling up the cover and several illustrations within. I parlayed that into a television production career. Over the course of the last twenty years, I advanced into animation, scripting, and then as a Producer/Director way out in the middle of the Pacific.

Physics has held a fascination for me. During college, I had enjoyed several "topics in physics" elective classes. They didn't relate to my degree in the least, but engaged a distinctive side of the brain all the same. I felt my intellectual draw could bridge a divide between their cerebral academic world and the one we everyday folk sip Starbucks in. Not to mention, if Neil deGrasse Tyson isn't inclined to step up, then I'm happy fill that vacuum.

We are laypeople. And if I can unravel these heady concepts into the realm of what's possible, then that has value for all of us.

Does that make me a UFOlogist?

Yes, and no. A UFOlogist is pretty much anyone who studies the topic and shares their research. Just like journalists, there isn't a UFOlogy certification. On top of that, those that seek to foster the stigma are more than pleased to slather pseudoscience over the term. In that respect, I'd not consider myself any sort of 'ologist, more of a layperson translator. I'm seeking to advance understanding. At the same time I'd like to push back against the disinformation that overwhelms the topic.

As a first time non-fiction author, I'm curious to see if this guidebook will spark something. Should it catch fire, I have plenty more in mind to continue this shared exploration. If you are curious about my other creative outlets, please check out the "Also by the Author" in the book's front, or my site at:

GOTurnerWrites.com

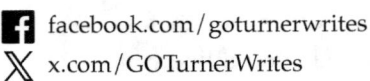 facebook.com/goturnerwrites
x.com/GOTurnerWrites

NOTES

5. BOB LAZAR'S SPORT MODEL

1. http://www.zamandayolculuk.com/ufonewtechnology.htm
2. September 22, 1990: "Ufos and the Alien Presence" by Michael Lindemann.
3. 1991: "The Lazar Tape," 40-minute VHS videotape
4. March, April 1990 (approximate time of interviews): "Alien Contact" by Timothy Good. Published by William Morrow and Company in 1991 & 1993.
5. October, 2022: YouTube Podcast Riccardo Storti - Bob Lazar's UFO Propulsion System Analysis

6. T. TOWNSEND BROWN'S GRAVITATOR

1. Force on an Asymmetric Capacitor https://arxiv.org/abs/physics/0211001

8. MARK MCCANDLISH'S FLUX LINER

1. https://en.wikipedia.org/wiki/Biefeld-Brown_effect
2. https://vault.fbi.gov/nikola-tesla

9. SALVATORE PAIS' NAVY PATENTS

1. https://bigthink.com/starts-with-a-bang/something-from-nothing/
2. https://www.sensonica.com/science/gertsenshtein-effect-and-how-it-works/

11. JACK SARFATTI'S LOW POWER WARP DRIVE

1. (36 minute mark) https://www.youtube.com/watch?v=ZAAq2i8Rw4k&t=626s

14. ORBS

1. https://nypost.com/2021/10/27/video-captures-pulsating-ufo-dropping-out-of-the-sky/
2. Afghanistan Orbs: https://www.youtube.com/watch?v=1MVDoA9GlnM&t=197s

16. CONJECTURE

1. https://www.youtube.com/watch?v=uQGhVBMFc1k

17. THREAT DIVISION

1. Wilson/Davis Memorandum - https://imgur.com/a/ggIFTfQ

www.ingramcontent.com/pod-product-compliance
Lightning Source LLC
Chambersburg PA
CBHW060524130626
46553CB00002B/645